Lecture Notes in Computer Science　10050

Commenced Publication in 1973
Founding and Former Series Editors:
Gerhard Goos, Juris Hartmanis, and Jan van Leeuwen

More information about this series at http://www.springer.com/series/7411

Marek Chrobak · Antonio Fernández Anta
Leszek Gąsieniec · Ralf Klasing (Eds.)

Algorithms for Sensor Systems

12th International Symposium on Algorithms and Experiments
for Wireless Sensor Networks, ALGOSENSORS 2016
Aarhus, Denmark, August 25–26, 2016
Revised Selected Papers

Springer

Editors
Marek Chrobak
University of California Riverside
Riverside, CA
USA

Antonio Fernández Anta
Institute IMDEA Networks
Leganés
Spain

Leszek Gąsieniec
University of Liverpool
Liverpool
UK

Ralf Klasing
Université de Bordeaux
Talence Cedex
France

ISSN 0302-9743   ISSN 1611-3349   (electronic)
Lecture Notes in Computer Science
ISBN 978-3-319-53057-4   ISBN 978-3-319-53058-1   (eBook)
DOI 10.1007/978-3-319-53058-1

Library of Congress Control Number: 2016963667

LNCS Sublibrary: SL5 – Computer Communication Networks and Telecommunications

Printed on acid-free paper

This Springer imprint is published by Springer Nature
The registered company is Springer International Publishing AG
The registered company address is: Gewerbestrasse 11, 6330 Cham, Switzerland

Preface

ALGOSENSORS, the International Symposium on Algorithms and Experiments for Wireless Sensor Networks, is an international forum dedicated to the algorithmic aspects of wireless networks, static or mobile. The 12th edition of ALGOSENSORS was held on August 25, 2016, in Aarhus, Denmark, as a part of the ALGO 2016 event.

Originally focused solely on sensor networks, ALGOSENSORS now covers more broadly algorithmic issues arising in all wireless networks of computational entities, including sensor networks, sensor-actuator networks, and systems of autonomous mobile robots. In particular, it focuses on the design and analysis of discrete and distributed algorithms, on models of computation and complexity, on experimental analysis, in the context of wireless networks, sensor networks, and robotic networks and on all foundational and algorithmic aspects of the research in these areas. This year papers were solicited for three tracks: Distributed and Mobile, Experiments and Applications, and Wireless and Geometry.

In response to the call for papers, 20 submissions were received, out of which nine papers were accepted after a rigorous reviewing process by the (joint) Program Committee, which involved at least three reviewers for each accepted paper. This volume contains the technical papers as well as an invited paper of the keynote talk by Fabian Kuhn (University of Freiburg).

We would like to thank all Program Committee members, as well as the external reviewers, for their fundamental contribution in selecting the best papers resulting in a strong program. We would also like to warmly thank the ALGO 2016 organizers for kindly accepting to co-locate ALGOSENSORS with some of the leading events on algorithms in Europe. Furthermore, we would like to thank the local ALGO 2016 Organizing Committee for their help regarding various administrative tasks, especially the local organizers Gerth Stølting Brodal (chair) and Trine Ji Holmgaard Jensen, as well as the Steering Committee chair, Sotiris Nikoletseas, for their help in ensuring a successful ALGOSENSORS 2016.

December 2016

Marek Chrobak
Antonio Fernández Anta
Leszek Gąsieniec
Ralf Klasing

Organization

Program Committee

Fernando Boavida	University of Coimbra, Portugal
Costas Busch	Louisiana State University, USA
Bogdan Chlebus	University of Colorado Denver, USA
Marek Chrobak	University of California Riverside, USA
Jurek Czyzowicz	Université du Quebec en Outaouais, Canada
Robert Elsässer	University of Salzburg, Austria
Antonio Fernadez Anta	Institute IMDEA Networks, Spain
Pierre Fraigniaud	CNRS and Université Paris Diderot, France
Leszek Gąsieniec	University of Liverpool, UK
Magnús M. Halldórsson	Reykjavik University, Iceland
Ralf Klasing	LaBRI - Université Bordeaux, France
Adrian Kosowski	IRIF (LIAFA)/Inria Paris, France
Evangelos Kranakis	Carleton University, Canada
Danny Krizanc	Wesleyan University, USA
Vincenzo Mancuso	Institute IMDEA Networks, Spain
Miguel A. Mosteiro	Kean University, USA
Tomasz Radzik	King's College London, UK
Gianluca Rizzo	HES SO Valais, Switzerland
Christian Scheideler	University of Paderborn, Germany
Christopher Thraves	Universidad de Concepción, Chile

Steering Committee

Sotiris Nikoletseas	University of Patras and CTI, Greece (Chair)
Josep Diaz	U.P. Catalunya, Spain
Magnús M. Halldórsson	Reykjavik University, Iceland
Bhaskar Krishnamachari	University of Southern California, USA
P.R. Kumar	Texas A&M University, USA

Additional Reviewers

Bampas, Evangelos
Clements, Wyatt
Di Stefano, Gabriele
Feldmann, Michael
Flocchini, Paola
Kolb, Christina
Krishnan, Hari
Lefevre, Jonas
Mallmann-Trenn, Frederik

Navarra, Alfredo
Pajak, Dominik
Scalosub, Gabriel
Setzer, Alexander
Shalom, Mordechai
Tamir, Tami
Tonoyan, Tigran
Wang, Haitao
Yu, Dongxiao

Invited Talk: Developing Robust Wireless Network Algorithms

Fabian Kuhn

Department of Computer Science, University of Freiburg, Freiburg, Germany
kuhn@cs.uni-freiburg.de

Over the last 30 years, we have seen a tremendous effort in the development of distributed algorithms and abstract models to deal with the characteristic properties of wireless communication networks. The models range from simple graph-based characterizations of interference to more accurate physical models such as the so-called signal-to-noise-and-interference (SINR) model. As different as the typically considered models are, most of them have one thing in common. Whether a node can successfully receive (and decode) a message is determined using some fixed, deterministic rule that depends only on the topology and structure of the actual network and on some additional model parameters.

While in classical wired networks, assuming reliable communication might be a reasonable abstraction, this seems much more problematic in a wireless network setting. The propagation of a wireless signal depends on many diverse environmental factors and it does not seem to be realistic to explicitly model all of these factors or to exactly measure the properties of the wireless communication channels. In addition, the environmental factors might change over time and there can also be additional independent sources of signal interference that cannot be predicted or controlled by the network. Further, wireless devices might also be mobile so that we not only have unreliable communication channels, but potentially even almost arbitrary dynamically changing network topologies. Because the classic abstract wireless communication models do not capture such unpredictable behavior, many existing radio network algorithms might only work in the idealized formal setting for which they were developed.

In the talk, we describe ways to develop more robust wireless network algorithms. In particular, we show that complex, unstable, unreliable, and also dynamic behavior of wireless communication networks can be modeled by adding a non-deterministic component to existing radio network models. As a result, any behavior which is too complex or impossible to predict or model explicitly is determined by an adversary. Clearly, such models might lead to less efficient algorithms. However, they also lead to more robust algorithms which tend to work under a much wider set of underlying assumptions. Very often, such models also lead to much simpler algorithms.

Contents

Contents

Multi-message Broadcast
in Dynamic Radio Networks

Mohamad Ahmadi[(✉)] and Fabian Kuhn

Department of Computer Science, University of Freiburg, Freiburg, Germany
{mahmadi,kuhn}@cs.uni-freiburg.de

Abstract. We continue the recent line of research studying information dissemination problems in adversarial dynamic radio networks. We
give two generic algorithms which allow to transform generalized version of single-message broadcast algorithms into multi-message broadcast
algorithms. Based on these generic algorithms, we obtain multi-message
broadcast algorithms for dynamic radio networks for a number of different dynamic network settings. For one of the modeling assumptions, our
algorithms are complemented by a lower bound which shows that the
upper bound is close to optimal.

1 Introduction

When developing algorithms for wireless networks, one has to deal with unique
challenges which are not or much less present in the context of wired networks.
All nodes share a single communication medium and whether a transmitted
signal can be received by a given node might therefore depend on the behavior
of all other nodes in the network. Moreover, the reception of a wireless signal
can be influenced by additional wireless devices, multi-path propagation effects,
other electric appliances, or also various additional environmental conditions,
see e.g., [25,35,37–39]. As a consequence, wireless connections often tend to be
unstable and unreliable. Moreover, wireless devices might be mobile, in which
case connectivity changes even when ignoring all the above effects. We therefore
believe that in order to study such dynamic and unpredictable communication
behavior, it is important to also study unreliable and dynamic variants of classic
wireless communication models such as, e.g., the classic radio network model
introduced in [6,10] or the more complex but also more realistic models like the
SINR model [23,33] or the affectance model [26].

In recent years, there has been a considerable effort in investigating basic
communication and distributed computation problems in radio network models
which exhibit adversarial dynamic, nondeterministic behavior. In [17], Clementi
et al. study the problem of broadcasting a single message in a synchronous
dynamic radio network where in each round a subset of the links might fail
adversarially. Communication is modeled using the standard graph-based radio
network model [6]. In each round, a node can either transmit or listen and

Research supported by ERC Grant No. 336495 (ACDC).

M. Chrobak et al. (Eds.): ALGOSENSORS 2016, LNCS 10050, pp. 1–15, 2017.
DOI: 10.1007/978-3-319-53058-1_1

a node u receives a message transmitted by a neighbor v in a given round r if and only if v is the only round-r neighbor of u transmitting in round r. The paper studies deterministic algorithms and it in particular shows that if D is the diameter of the fault-free part of the network, deterministic single-message broadcast requires time $\Theta(Dn)$, where n is the number of nodes of the network. In [15], Clementi et al. study an even more dynamic network model where the network topology can completely change from round to round. It is in particular shown that the single-message broadcast problem can be solved in time $O(n^2/\log n)$ by a simple randomized algorithm where in each round, each node knowing the broadcast message, transmits it with probability $\ln(n)/n$. It is also shown that the asymptotic time complexity of this algorithm is optimal. A similar model to the one in [17] has been studied in [29]. In the *dual graph* model of [29], it is assumed that the set of nodes is fixed and the set of edges consists of two parts, a connected set of reliable edges and a set of unreliable edges. In each round, the communication graph consists of all reliable and an arbitrary (adversarially chosen) subset of the unreliable edges. Among other results, it is shown that there is a randomized algorithm which achieves single-message broadcast in time $O(n \log^2 n)$. The algorithm in [29] works in the presence of a *strongly adaptive* adversary which determines the set of edges of a given round r after the distributed algorithm decides what messages are transmitted by which nodes in round r. In [21], the same problem is considered for weaker adversaries. A *weakly adaptive* adversary has to determine the topology of round r before a randomized distributed algorithm determines the randomness of round r and an *oblivious adversary* has to determine the whole sequence of network topologies at the very beginning of the execution of a distributed algorithm. Additional problems in the dual graph model of [21,29] have also been studied in [9,22,32].

The dynamic network models of the above papers can (mostly) be seen as the extreme cases of the *T-interval connected* dynamic graph model of [30]. For a positive integer T, a dynamic network is called T-interval connected, if for any time interval of T consecutive rounds the graph which is induced by the set of edges present throughout the whole time interval is connected. Hence, for $T = 1$, the network graph has to be connected in every round, whereas for $T = \infty$, we obtain the dual graph model. In [1], the single-message broadcast problem has been studied for general T-interval connected dynamic radio networks and for a more fine-grained adversary definition. More specifically, for an integer $\tau \geq 0$, an adversary is called τ-oblivious if the network topology of round r is determined based on the knowledge of all random decisions of the distributed algorithm of rounds $1, \ldots, r - \tau$. Hence, an oblivious adversary is ∞-oblivious, a weakly adaptive adversary is 1-oblivious, and a strongly adaptive adversary is 0-oblivious.

Additional Related Work. Information dissemination and specifically broadcasting is a fundamental problem in networks and therefore there exists a rich literature on distributed algorithm to solve broadcast in various communication settings. In particular, the problem is well-studied in static wireless networks (see, e.g., [5,6,10,11,16] and many more). More recently, the multi-message problem has been studied quite extensively in the context of wireless networks, see e.g., [12–14,19,20,24,28].

We also note that several dynamic network models similar to the one in [30] have been studied prior to [30], for example in [4,7,8,15,36]. For a recent survey, we also refer to [31]. In addition to the works already mentioned, information dissemination and other problems in faulty and dynamic radio network models have also been studied in, e.g., [3,17,27].

1.1   Contributions

In the present work, we extend [1] and more generally the above line of research and we study the multi-message broadcast problem in T-interval connected dynamic radio network against a τ-oblivious adversary. For some $s \geq 1$, we assume that there are s broadcast messages, each initially given to a single source node. A randomized distributed algorithm solves the s-multi-message broadcast problem if with high probability (w.h.p.), it disseminates all s broadcast messages to all nodes of the network. We say that the communication capacity of a network is $c \geq 1$ if every message sent by a node can contain up to c different broadcast messages.

Upper Bounds. Most of our upper bounds are achieved by *store-and-forward* algorithms, i.e., by algorithms which treat the broadcast messages as single units of data that cannot be altered or split into several pieces. All our store-and-forward protocols are based on two generic algorithms which allow to turn more general variants of single-message broadcast algorithms into multi-message broadcast algorithms. In addition, Theorem 4 proves an upper bound that can be achieved using random linear network coding. When dealing with linear network coding algorithms, we use the common convention and disregard any overhead created by the message headers describing the linear combination of broadcast messages sent in the message (see, e.g., [18]).[1]

Theorem 1. *Assume that we are given an ∞-interval connected dynamic n-node network controlled by a 0-oblivious adversary. Using store-and-forward algorithms, for communication capacity $c = 1$, s-multi-message broadcast can be solved, w.h.p., in time $O(ns \log^2 n)$, whereas for arbitrary $c \geq 1$, it can be solved, w.h.p., in time $O((1 + \frac{s}{c})n \log^4 n)$.*

Theorem 2. *Consider the s-multi-message broadcast problem in 1-interval connected dynamic n-node networks controlled by a 1-oblivious adversary. Applying store-and-forward algorithms, for communication capacity $c = 1$, the problem can be solved, w.h.p., in time $O((1 + \frac{s}{\log n})n^2)$, and for arbitrary $c \geq 1$, it can be solved, w.h.p., in time $O((1 + \frac{s}{c})n^2 \log n)$.*

---

[1] Note that this assumption is reasonable as long as the number of broadcast messages which are combined with each other is at most the length of a single broadcast message (in bits).

Theorem 3. *Let $T \geq 1$ and $\tau \geq 1$ be positive integer parameters. In T-interval connected dynamic n-node networks controlled by a τ-oblivious adversary, for communication capacity $c = 1$, the s-multi-message broadcast problem can be solved, w.h.p., in time*

$$O\left(\left(1 + \frac{n}{min\{\tau, T\}}\right) \cdot (s + \log n) \cdot n \cdot \log^3 n\right),$$

and for an arbitrary c, it can be solved, w.h.p., in time

$$O\left(\left(1 + \frac{n}{min\{\tau, T\}}\right) \cdot \frac{ns}{c} \cdot \log^4 n\right).$$

Theorem 4. *Using linear network coding, in 1-interval connected dynamic networks with communication capacity 1 and a 1-oblivious adversary, s-multi-message broadcast can be solved in time $O(n^2 + ns)$, w.h.p.*

Lower Bound

Theorem 5. *In ∞-interval connected dynamic networks with communication capacity $c \geq 1$ and a 0-oblivious adversary, any s-multi-message broadcast algorithm requires at least time $\Omega(ns/c)$, even when using network coding. Further, there is a constant-diameter ∞-interval connected network with communication capacity 1 such that any store-and-forward algorithm requires at least $\Omega((ns - s^2)/c)$ rounds to solve s-multi-message broadcast against a 0-oblivious adversary.*

2   Model and Problem Definition

Dynamic Networks. We model dynamic radio networks using the synchronous dynamic network model of [30].[2] A dynamic network is represented by a fixed set of nodes V of size n and a sequence of undirected graphs $\langle G_1, G_2, \ldots \rangle$, where $G_i = (V, E_i)$ is the communication graph in round i. Hence, while the set of nodes remains the same throughout an execution, the set of edges can potentially change from round to round. A dynamic graph $\langle G_1, G_2, \ldots \rangle$ is called *T-interval connected* for an integer parameter $T \geq 1$ if the graph

$$\overline{G}_{r,T} = (V, \overline{E}_{r,T}), \quad \text{where } \overline{E}_{r,T} := \bigcap_{r'=r}^{r+T-1} E_{r'}$$

is connected for all $r \geq 1$.

Communication Model. We define an n-node distributed algorithm $\mathcal{A}$ as a collection of n processes which are assigned to the nodes of an n-node network. Thus, at the beginning of an execution, a bijection from V to $\mathcal{A}$ is defined by

---

[2] We note that many other, similar dynamic network models have appeared in the literature (cf. Sect. 1).

an adversary. In the following, we will use "node" to refer to a node $u \in V$ and to the process running at node u. We assume that each node has a unique ID of $O(\log n)$ bits. In each round of an execution, each node u decides to either transmit a message or listen to the wireless channel. When a node decides to transmit a message in round r, the message reaches all of its neighbors in G_r. A node u in round r successfully receives a message from a neighbor v if and only if v is the only neighbor of u in G_r transmitting in round r. Otherwise, if zero or multiple messages reach a node u, u receives silence, i.e., nodes cannot detect collisions.

Adversary. We assume that the dynamic graph $\langle G_1, G_2, \dots \rangle$ is determined by an adversary. Classically, in this context, three types of adversaries have been considered (see, e.g., [21]). An *oblivious* adversary has to determine the whole sequence of graphs at the beginning of an execution, *independently* of any randomness used in the algorithm. An *adaptive* adversary can construct the graph of round r depending on the history and thus in particular the randomness up to round r. Typically, two different adaptive adversaries are considered. A *strongly adaptive* adversary can choose graph G_r dependent on the history up to round r including the randomness of the algorithm in round r, whereas a *weakly adaptive* adversary can only use the randomness up to round $r-1$ to determine G_r. In the present paper, we use a more fine-grained adversary definition which was in this form introduced in [1]. For an integer $\tau \geq 0$, we call an adversary τ-*oblivious* if for any $r \geq 1$, the adversary constructs G_r based on the knowledge of the algorithm description and the algorithm's random choices of the first $r - \tau$ rounds. Note that the three classic adversaries described above are all special cases of a τ-oblivious adversary, where $\tau = \infty$ corresponds to an oblivious adversary, $\tau = 0$ to a strongly adaptive adversary and $\tau = 1$ to a weakly adaptive adversary.

Multi-message Broadcast Problem. For some positive integers s and B, we define the s-multi-message broadcast problem as follows. We assume that there are s distinct *broadcast messages* of size B bits, each of which is given to some node in the network, called a *source node*. Then the problem requires the dissemination of these s broadcast messages to all nodes in the network. It can be solved by two types of algorithms; *store-and-forward algorithms* and *network coding algorithms*. In store-and-forward algorithms, each broadcast message is considered as a black box, and nodes can only store and forward them. In contrast to store-and-forward algorithms, in network coding algorithms, each node can send a message which can be any function of the messages it has received so far. In the following we define two types of problems and later we show that a procedure that can solve each of these problems can be used as a subroutine for solving the multi-message broadcast problem.

Communication Capacity. To solve the multi-message broadcast problem we consider a restriction on the amount of data which can be transmitted in a single message of a distributed protocol. The *communication capacity* $c \geq 1$ is defined as the maximum number of broadcast messages that can be sent in a single message of a distributed algorithm. In addition, an algorithm can send some additional control information of the same (asymptotic) size, i.e., an algorithm is allowed to send messages of $O(cB)$ bits.

Limited Single-Message Broadcast Problem. Let us assume that there is a single broadcast message initially given to some node in an n-node network. For some integer parameter $k \leq n$, k-limited broadcast problem requires the successful receipt of the broadcast message by at least k nodes in the network with probability at least $1/2$.

Concurrency-Resistant Single-Message Broadcast Problem. In this problem we assume that there can be 0, 1, or more source nodes in the network, each of which is given a broadcast message. If there exists only one source node, then its broadcast message is required to be successfully received by all nodes in the network, and thus this execution is considered a successful broadcast. Otherwise, we have an unsuccessful broadcast, where there are no source nodes or more than one source nodes. The problem requires that all nodes detect whether the broadcast was successful or not by the end of the execution. This broadcast with success detection is actually simulating a single-hop communication network with collision detection, where if only one node broadcasts in a round, all nodes receive the message and otherwise all nodes detect collision/silence.

3   Multi-message Broadcast Algorithms

In this section we present upper bounds for the multi-message broadcast problem in different scenarios depending on the communication capacity, the interval connectivity of the communication network, the adversary strength, and also the ability to use network coding for disseminating information. We start by describing *generic techniques* for broadcasting multiple messages in dynamic radio networks. Due to lack of space, most proofs, some algorithm descriptions, and also the discussion of how to use network coding to solve multi-message broadcast in dynamic radio networks appear in the full version of this paper [2].

3.1   Generic Algorithm for Large Communication Capacity c

We start by describing a generic method for coping with a general communication capacity parameter c (see Sect. 2). If we want to design store-and-forward algorithms which exploit the fact that in a given time slot a node can transmit $c \gg 1$ source messages to its neighbors, we have to deal with the problem that initially, all broadcast messages might start at distinct source nodes. In order to collect sufficiently many broadcast messages at single nodes and to nevertheless avoid too many redundant retransmission of the same broadcast message by several nodes, we adapt a technique introduced by Chrobak et al. in [14]. Their algorithm runs in phases and it is based on iterative applications of a k-limited single-message broadcast routine for exponentially growing values of k. In each phase, for each broadcast message $\mathcal{M}$, the minimum number of nodes which know $\mathcal{M}$ doubles. Typically, the maximum time for reaching at least k nodes with a single-message broadcast algorithm grows linearly in k. If k is doubled in each phase, the time for each k-limited single-message broadcast instance therefore also doubles. However, since in each phase, the number of nodes which know each source message $\mathcal{M}$ at the beginning of a phase also doubles, the number of source nodes from which we need to start a k-limited single-message broadcast instance gets divided by 2 and overall, the time complexity of each phase will be about the same.

Distributed Coupon Collection. Formally, in [14] this idea is modeled by a *distributed coupon collection* process which we generalize here. The distributed coupon collection problem is characterized by four positive integer parameters n, s, ℓ, and c (in [14], the parameters s and c are both equal to n). There are n bins and s distinct coupons (in the application, the bins will correspond to the nodes of the communication network and the distinct coupons will be the broadcast messages). There are at least ℓ copies of each coupon and the at least $s\ell$ coupons are distributed among the n bins such that each bin contains at most one copy of each of the s distinct coupons. The coupon collection proceeds in discrete time steps. In each time step, a bin is chosen uniformly at random. If a bin is chosen, it is opened with probability $1/2$. If the bin is opened, at most c coupons of it are collected as follows. If the bin has at most c coupons, all coupons of the bin are collected, otherwise a randomly chosen subset of size c of the coupons in the bin is collected. The coupon collection ends as soon as each of the s distinct coupons has been collected at least once. The following lemma upper bounds the total number of time steps needed to collect all s coupons.

Lemma 1 (Distributed Coupon Collection). *With high probability, the described distributed coupon collection process ends after $O\left(\frac{ns}{c\ell} \cdot \log(n+s)\right)$ steps.*

Coupon-Collection-Based Generic Multi-message Broadcast Algorithm. We now discuss how to use the above abstract distributed coupon collection process to efficiently broadcast multiple messages in a dynamic radio network. Our algorithm is a generalization of the idea of Chrobak et al. [14]. The algorithm consists of $\lceil \log_2 n \rceil$ phases. By the end of phase i, for each broadcast message $\mathcal{M}$, we want to guarantee that at least $\ell_i := 2^i$ nodes know $\mathcal{M}$. We can therefore assume that at the beginning of a phase, each source message $\mathcal{M}$ is known by at least $\ell_{i-1} = 2^{i-1}$ nodes (note that this is trivially true for $i = 1$). We can achieve that each broadcast message is known by ℓ_i nodes by running sufficiently many instances of ℓ_i-limited broadcast such that each source message $\mathcal{M}$ is disseminated in $O(\log n)$ of these ℓ_i-limited broadcast instances. The details of the algorithm are given by Algorithm 1. In the pseudocode of Algorithm 1, $\alpha > 0$ is a constant which is chosen sufficiently large.

The following lemma shows that if the cost of k-limited broadcast depends at most linearly on k, Algorithm 1 achieves s-multi-message broadcast in essentially the time needed to perform s/c single-message broadcasts (i.e., n-limited broadcasts).

Algorithm 1. Generic Multi-message Broadcast Algorithm Based on Distributed Coupon Collection.

1: **for each** $v \in V$ **do** $S_v \leftarrow$ set of broadcast messages known by v
2: **for** $i \leftarrow 1$ **to** $\lceil \log_2 n \rceil$ **do**
3:     $\ell_i \leftarrow 2^i$
4:     **for** $j \leftarrow 1$ **to** $\alpha \cdot \frac{ns}{c\ell_i} \cdot \ln n$ **do**
5:         **for each** $v \in V$ **do**
6:             (independently) mark v with probability $1/n$
7:             **if** v is marked **then**
8:                 $R_v \leftarrow$ random subset of S_v of size $\min\{|S_v|, c\}$
9:                 v initiates ℓ_i-limited broadcast with message R_v

Lemma 2. *Let $t(k, n)$ be the time needed to run one instance of k-limited broadcast in an n-node network. Then, when using a sufficiently large constant $\alpha > 0$ in Algorithm 1, the algorithm w.h.p. solves s-multi-message broadcast in time*

$$O\left(\sum_{i=1}^{\lceil \log_2 n \rceil} \frac{ns}{c2^i} \cdot t(2^i, n) \cdot \log n \right).$$

Typically, the time to run one instance of k-limited broadcast depends linearly on k. The following corollary simplifies the statement of Lemma 2 for this particular case.

Corollary 1. *Assume that the time for running one instance of k-limited broadcast in an n-node network is given by $t(k, n) \leq k \cdot t(n)$. Then, when using a sufficiently large constant $\alpha > 0$ in Algorithm 1, the algorithm w.h.p. solves s-multi-message broadcast in time*

$$O\left(t(n) \cdot \frac{ns}{c} \cdot \log^2 n \right).$$

Proof. Follows directly from Lemma 2.

3.2 Generic Algorithm for Constant Communication Capacity c

We next describe a more efficient generic algorithm for the case $c = 1$ (or any constant c). Hence, we assume that each message sent by the algorithm can only contain a single broadcast message. In this case, we do not need to care about collecting different broadcast messages at a single node and Algorithm 1 is therefore too costly. Instead, we use an algorithm which is based on a more standard single-message broadcast algorithm. In the following, we assume that for some given setting we are given a *concurrency-resistant* single-message broadcast algorithm $\mathcal{B}$. Recall that if there exists only one broadcast message while running $\mathcal{B}$, all nodes receive the broadcast message and $\mathcal{B}$ returns 1, otherwise it returns 0. We use this algorithm as a subroutine in designing the generic multi-message broadcast algorithm. We assume that initially, the number of broadcast

Algorithm 2. Generic Multi-message Broadcast Algorithm Based on Concurrency-Resistant Single-Message Broadcast Algorithm.

1: **for each** $v \in V$ **do** $S_v \leftarrow$ set of broadcast messages given initially to v
2: $x \leftarrow s$
3: **while** $x \neq 0$ **do**
4:     **for each** $v \in V$ **do**
5:         **if** $S_v \neq \emptyset$ **then**
6:             (independently) mark each of the broadcast messages in S_v with probability $1/x$
7:         **if** v has a marked broadcast message **then**
8:             v initiates $\mathcal{B}$ with its marked broadcast messages
9:         **if** $\mathcal{B}$ returns 1 **then**
10:             $x \leftarrow x - 1$
11:             remove the broadcast message of node u which is delivered to all nodes from S_u
12:             unmark all broadcast messages

messages s is known. The generic algorithm runs in phases and in each phase we run one instance of $\mathcal{B}$. Note that if the instance returns 1, all nodes know the broadcast message which has been disseminated to all nodes. Therefore at all times, all nodes know how many messages still need to be broadcast. If at the beginning of a phase, there are $x \leq s$ broadcast messages which still need to be broadcast, for each broadcast message $\mathcal{M}$, the source node of $\mathcal{M}$ decides to start an instance of $\mathcal{B}$ with broadcast message $\mathcal{M}$ with probability $1/x$.

Lemma 3. *If $\mathcal{B}$ is a concurrency-resistant single-message broadcast algorithm with running time $t(n)$, then Algorithm 2 solves the s-multi-message broadcast problem in time $O\big((s + \log n)t(n)\big)$.*

In the sequel, we apply this generic broadcast algorithm to solve the multi-message broadcast problem in three different settings.

3.3 Application of the Generic Algorithms in Different Settings

In this section we intend to show how to apply the generic multi-message broadcast techniques introduced in Sects. 3.1 and 3.2 in different settings. In the following we consider the problem in networks with different interval connectivity T. In each of these settings we investigate the two cases of large and constant communication capacities separately. For the first case the generic algorithm requires the existence of a k-limited single-message broadcast algorithm and in the second case the generic algorithm requires the existence of a concurrency-resistant single-message broadcast algorithm. We therefore need to show that the existing single-message broadcast algorithms in the considered dynamic radio network settings can be turned into k-limited and concurrency-resistant variants of these algorithms.

(Setting I) ∞-Interval Connected Dynamic Networks. We consider the s-multi-message broadcast problem in an ∞-interval connected dynamic network against a 0-oblivious adversary. To obtain a k-limited and a concurrency-resistant single-message broadcast algorithm, we adapt the harmonic broadcast algorithm introduced in [29]. In the harmonic broadcast algorithm, in the first round, the source node transmits its broadcast message to its neighbors. From the second round on, any node which receives the broadcast message in round r_v, transmits the message in any round $r > r_v$ with probability $p_v(r)$, given by

$$p_v(r) := \frac{1}{1 + \lfloor \frac{r-r_v-1}{T} \rfloor},$$

where $T = \Theta(\log n)$ is a parameter. That is, for the first T rounds immediately after some node receives the message, it transmits the message with probability 1, for the next T rounds it transmits with probability $1/2$, then with probability $1/3$ and so on. In the full version of the paper [2], we show that when running the harmonic broadcast algorithm from [29] for $4kT(\ln n + 1)$ rounds, it solves the k-limited single-message broadcast problem. The second statement of Theorem 1 then follows directly from Corollary 1.

In order to obtain a concurrency-resistant broadcast algorithm, we run the (full) harmonic broadcast algorithm twice. First, the algorithm is run with the given broadcast messages. If there are at least 2 different broadcast messages, one can show that w.h.p., at least one node receives two different broadcast messages. Every node which receives at least two different broadcast messages in the first instance of the harmonic algorithm, broadcasts $\perp$ in the second instance. It can also be shown that the harmonic broadcast algorithm of [29] also works if several nodes try to broadcast the same message. Hence, if a node receives a message in the first instance of the harmonic algorithm and receives nothing in the second instance, the concurrency-resistant broadcast was successful. If a node receives nothing in the first instance or $\perp$ in the second instance, the concurrency-resistant broadcast algorithm did not succeed. For details, we refer to the full version of this paper [2].

(Setting II) 1-Interval Connected Dynamic Networks. For 1-interval connected networks, we assume that the adversary is 1-oblivious. Note that in [1], it is shown that even the single-message problem cannot be solved with a 0-oblivious adversary in 1-interval connected networks. We adapt the homogenous algorithm by Clementi et al. [15] to obtain a k-limited and a concurrency-resistant single-message broadcast procedure. The single-message broadcast algorithm of [15] is a simple randomized algorithm where every node that knows the broadcast message, broadcasts the message with probability $\ln(n)/n$. In [15], it is shown that in every round, the probability that a node knowing the broadcast message succeeds in sending it to a neighboring node which does not know the message is at least $\ln(n)/n$. In order to have a k-limited broadcast algorithm which succeeds in reaching at least k nodes with probability at least $1/2$, we therefore need to run the algorithm of [15] for $\Theta(kn/\ln n)$ rounds. Hence, the second statement of Theorem 2 follows directly from Corollary 1.

In order to obtain a concurrency-resistant algorithm, we can first observe that the algorithm of [15] also works if several nodes try to broadcast the same message. Hence, in order to use a similar idea as in the ∞-interval connected case, we have to make sure that in the case of multiple broadcast messages, some node detects that there are at least two messages. Using the same analysis as for a single broadcast message, one can see that after $O(n^2/\log n)$ rounds, every node knows at least one of the broadcast messages. Our next goal is to have at least one node which knows at least two different broadcast messages. As long as this is not the case, in every round, there is at least one edge connecting two nodes u and v which know different broadcast messages. For the next $O(n\log n)$ rounds, each node now transmits the broadcast message it knows with probability $1/n$ (assuming a node knows exactly one broadcast message). The probability that either u or v decide to broadcast their message and no other node in the network decides to broadcast is at least $2/cn$ and thus after $O(n\log n)$ rounds, w.h.p., there is at least one node which knows two broadcast messages.

Now, we can get a concurrency-resistant single-message broadcast algorithm in the same way as for ∞-interval connected graphs. We run the algorithm of [15] again such that each node which knows at least two different broadcast messages after the first phase, broadcasts $\perp$. The time complexity of the concurrency-resistant broadcast algorithm is $O(n^2/\log n)$ and therefore the first statement of Theorem 2 now directly follows from Lemma 3.

(Setting III) **T-Interval Connected Dynamic Networks.** For arbitrary T-interval connected graphs, we adapt the algorithm of [1]. We assume that the adversary is τ-oblivious and we define $\psi := \min\{\tau, T\}$. We assume that $\psi = \Omega(\log^2 n)$, note that otherwise, Theorem 3 already follows from the statement of Theorem 2 for $T = 1$. We also assume that $\psi = O(n)$, as otherwise Theorem 3 follows from Theorem 1 for $T = \infty$. The algorithm of [1] runs in phases of length ψ. The progress in a single phase is analyzed in the proof of Theorem 1. A phase of length ψ is called successful if at least an $\Omega(\psi/(n\log^2 n))$-fraction of all uninformed nodes are informed and it is shown that a phase is successful with probability at least $\psi/(8en)$.

Assume that we want to obtain a k-limited single-message broadcast algorithm. If $k \leq n/2$, as long as fewer than k nodes are informed, the number of uninformed nodes is at least $n/2$. Hence, in this case, in a successful phase, at least $\Omega(\psi/\log^2 n)$ nodes are newly informed. Hence, to inform $k \leq n/2$ nodes, we need $O(k\log^2(n)/\psi)$ successful phases and we thus need $O(kn\log^2(n)/\psi^2)$ phases in total to inform k nodes with probability at least $1/2$. The number of rounds to solve k-limited single-message broadcast is therefore $O(kn\log^2(n)/\psi) = O(kn\log^2(n)/\min\{\tau, T\})$. For $k > n/2$, we run the complete single-message broadcast algorithm of [1] and obtain a time complexity of $O(n^2\log^3(n)/\min\{\tau, T\})$. The second statement of Theorem 3 now follows directly from Lemma 2.

To obtain a concurrency-resistant algorithm we use the same approach as in the 1-interval connected case. We first run the algorithm of [1]. From the description and analysis of the algorithm is not hard to see that the algorithm also works if several source nodes start the algorithm with the same broadcast message. In addition, the transmission behavior of a node does not depend on the content of the broadcast message and thus, if there are 2 or more broadcast messages, one can easily see that at the end of the algorithm, either there exists at least one node which knows at least two different broadcast messages or each node knows exactly one message. If every node knows exactly one broadcast message, we can create one node which knows at least two different broadcast messages in $O(n \log n)$ rounds in the same way as in the 1-interval connected case. The concurrency-resistant algorithm is then completed by running the algorithm of [1] once more, where every node which knows at least two different broadcast messages, uses the algorithm to broadcast $\bot$ to all nodes. The total time complexity of the algorithm is $O(n^2 \log^3(n) / \min\{\tau, T\} + n \log n)$ and therefore the first statement of Theorem 3 follows from Lemma 3.

4   Multi-message Broadcast Lower Bound

In this section we give a lower bound for solving the multi-message broadcast problem in an ∞-interval connected network controlled by a 0-oblivious adversary.

Recall that in an ∞-interval connected dynamic network, there is a static connected spanning subgraph which is present every round. In the following, we refer to this graph as the stable subgraph. We first prove a simple lower bound for the case where the stable subgraph has a non-constant diameter. With a more involved argument, we then extend the lower bound to the case where the stable subgraph has a constant diameter.

Lemma 4. *Assume that G is an n-node graph with maximum degree $\Delta = O(1)$. If G is the stable subgraph of an ∞-interval connected dynamic radio network, with a 0-oblivious adversary, every s-multi-message broadcast algorithms requires at least $\Omega(ns/c)$ rounds, where $c \geq 1$ is the communication capacity of the network.*

Proof. Recall that a 0-oblivious adversary can construct the communication graph of a given round r after the random decisions of all nodes in round r. A 0-oblivious adversary therefore in particular knows which nodes are transmitting in round r before determining the graph of round r. Given an s-multi-message broadcast algorithm $\mathcal{A}$, an 0-oblivious adversary constructs the sequence of communication graphs as follows. In every round in which 2 or more nodes decide to transmit, the communication graph is a complete graph. In all other rounds, the communication graph is only the stable graph G. Hence, in rounds with 2 or more nodes transmitting, all n nodes will experience a collision and therefore the algorithm cannot make any progress. In rounds where 0 nodes are transmitting, there clearly also cannot be any progress. In rounds where exactly one node v is transmitting, the message of v only reaches its at most $\Delta = O(1)$ neighbors in G. Because we have s broadcast messages of B bits and because each broadcast message only has one source node, over the whole algorithm, the nodes in total need to learn $\Theta(nsB)$ bits of information. As every message sent by the algorithm can contain only $O(cB)$, in each round, the total number of bits learned by any node is also at most $O(cB)$. The lemma therefore follows.   □

To prove the lower bound for constant-diameter stable subgraphs, we use the hitting game technique introduced by Newport in [34]. This is a general technique to prove lower bounds for solving various communication problems in radio networks. Using this technique, one first defines an appropriate combinatorial game with respect to the problem such that a lower bound for the game can be proved directly. It is shown that an efficient solution for the radio network problem helps a player to win the game efficiently. Consequently, the game's lower bound can be leveraged to the problem's lower bound.

For the sake of proving this lower bound, we define a combinatorial game called (α, β)-hitting game. Assuming the existence of a distributed algorithm $\mathcal{A}$ which solves multi-message broadcast in the desired setting, we will show that a player can simulate the execution of $\mathcal{A}$ in an ∞-interval connected dynamic network called the *target network*. Then, the player uses the transmitting behavior of the nodes in the target network, while running $\mathcal{A}$, to win the game efficiently. For space reasons, the detailed lower bound proof appears in the full version of this paper [2].

5 Conclusions

In the paper, we studied multi-message broadcast in T-interval connected radio networks with different adaptive adversaries. In all considered cases, we shows that if c broadcast messages can be packed into a single message (of the algorithm), s broadcast messages can essentially be broadcast in s/c times the time required to broadcast a single message. In one case (∞-interval connected dynamic networks with a 0-oblivious adversary), we also showed that up to logarithmic factors, our algorithm is optimal. Note that using techniques from [1,21], at the cost of one logarithmic factor, this lower bound can also be adapted to work in the presence of a 1-oblivious adversary.

A multi-message broadcast time which is roughly s/c times as large as the time for broadcasting a single message seems not very spectacular. Such an algorithm essentially always runs just one single-message broadcast algorithm at each point in time (where for $c > 1$, the algorithm each time broadcasts a collection of messages). However, we believe that it will be interesting to see whether the time complexity can be significantly improved in any of the adversarial dynamic network settings considered in this paper. When using store-and-forward algorithms, such an improvement would imply that the algorithm can use some form of pipelining in an efficient manner. I might also be interesting to study somewhat weaker (adversarial) dynamic network models which allow some pipelining when broadcasting multiple messages.

References

1. Ahmadi, M., Ghodselahi, A., Kuhn, F., Molla, A.R.: The cost of global broadcast in dynamic radio networks. In: Proceedings of the 19th International Conference on Principles of Distributed Systems (OPODIS) (2015)

2. Ahmadi, M., Kuhn, F.: Multi-message broadcast in dynamic radio networks. CoRR, abs/1610.02931 (2016)
3. Anta, A.F., Milani, A., Mosteiro, M.A., Zaks, S.: Opportunistic information dissemination in mobile ad-hoc networks: the profit of global synchrony. Distrib. Comput. **25**(4), 279–296 (2012)
4. Avin, C., Koucký, M., Lotker, Z.: How to explore a fast-changing world (cover time of a simple random walk on evolving graphs). In: Proceedings of the 5th Colloquium on Automata, Languages and Programming (ICALP) (2008)
5. Bar-Yehuda, R., Goldreich, O., Itai, A.: Efficient emulation of single-hop radio network with collision detection on multi-hop radio network with no collision detection. Distrib. Comput. **5**(2), 67–71 (1991)
6. Bar-Yehuda, R., Goldreich, O., Itai, A.: On the time-complexity of broadcast in multi-hop radio networks: an exponential gap between determinism and randomization. Comput. Syst. Sci. **45**(1), 104–126 (1992)
7. Baumann, H., Crescenzi, P., Fraigniaud, P.: Parsimonious flooding in dynamic graphs. In: Proceedings of the 28th ACM Symposium on Principles of Distributed Computing (PODC) (2009)
8. Casteigts, A., Flocchini, P., Quattrociocchi, W., Santoro, N.: Time-varying graphs and dynamic networks. In: Proceedings of the 10th International Conference on Ad-hoc, Mobile, and Wireless Networks (2011)
9. Censor-Hillel, K., Gilbert, S., Kuhn, F., Lynch, N., Newport, C.: Structuring unreliable radio networks. Distrib. Comput. **27**(1), 1–19 (2014)
10. Chlamtac, I., Kutten, S.: On broadcasting in radio networks-problem analysis and protocol design. IEEE Trans. Commun. **33**(12), 1240–1246 (1985)
11. Chlebus, B.S., Gasieniec, L., Ostlin, A., Robson, J.M.: Deterministic radio broadcasting. In: Proceedings of the 27th International Colloquium on Automata, Languages, and Programming (ICALP) (2000)
12. Chlebus, B.S., Kowalski, D.R., Radzik, T.: Many-to-many communication in radio networks. Algorithmica **54**(1), 118–139 (2009)
13. Christersson, M., Gasieniec, L., Lingas, A.: Gossiping with bounded size messages in ad hoc radio networks. In: Proceedings of the 29th International Colloquium on Automata, Languages and Programming (ICALP) (2002)
14. Chrobak, M., Gasieniec, L., Rytter, W.: A randomized algorithm for gossiping in radio networks. Networks **43**(2), 119–124 (2004)
15. Clementi, A., Monti, A., Pasquale, F., Silvestri, R.: Broadcasting in dynamic radio networks. Comput. Syst. Sci. **75**(4), 213–230 (2009)
16. Clementi, A., Monti, A., Silvestri, R.: Selective families, superimposed codes, and broadcasting on unknown radio networks. In: Proceedings of the ACM-SIAM Symposium on Discrete Algorithms (SODA) (2001)
17. Clementi, A., Monti, A., Silvestri, R.: Round robin is optimal for fault-tolerant broadcasting on wireless networks. Parallel Distrib. Comput. **64**(1), 89–96 (2004)
18. Deb, S., Médard, M., Choute, C.: Algebraic gossip: a network coding approach to optimal multiple rumor mongering. IEEE/ACM Trans. Inf. Theory **52**(6), 2486–2507 (2006)
19. Gasieniec, L., Kranakis, E., Pelc, A., Xin, Q.: Deterministic m2m multicast in radio networks. Theor. Comput. Sci. **362**(1–3), 196–206 (2006)
20. Ghaffari, M., Kantor, E., Lynch, N., Newport, C.: Multi-message broadcast with abstract mac layers and unreliable links. In: Proceedings of the 2014 ACM Symposium on Principles of Distributed Computing (PODC) (2014)

21. Ghaffari, M., Lynch, N., Newport, C.: The cost of radio network broadcast for different models of unreliable links. In: Proceedings of the 32nd Symposium on Principles of Distributed Computing (PODC) (2013)
22. Ghaffari, M., Newport, C.: Leader election in unreliable radio networks. In: Proceedings of the 43rd International Colloquium on Automata, Languages and Programming (ICALP) (2016)
23. Gupta, P., Kumar, P.R.: The capacity of wireless networks. IEEE Trans. Inf. Theory **46**(2), 388–404 (2000)
24. Khabbazian, M., Kowalski, D.R.: Time-efficient randomized multiple-message broadcast in radio networks. In: Proceedings of the 30th ACM SIGACT-SIGOPS Symposium on Principles of Distributed Computing (PODC)
25. Kim, K.-H., Shin, K.G.: On accurate measurement of link quality in multi-hop wireless mesh networks. In: Proceedings of the 12th International Conference on Mobile Computing and Networking (MOBICOM) (2006)
26. Kowalski, D.R., Mosteiro, M.A., Rouse, T.: Dynamic multiple-message broadcast: bounding throughput in the affectance model. In: Proceedings of the 10th ACM International Workshop on Foundations of Mobile Computing (2014)
27. Kranakis, E., Krizanc, D., Pelc, A.: Fault-tolerant broadcasting in radio networks. Algorithms **39**(1), 47–67 (2001)
28. Kuhn, F., Lynch, N., Newport, C.: The abstract mac layer. Distrib. Comput. **24**(3), 187–206 (2011)
29. Kuhn, F., Lynch, N., Newport, C., Oshman, R., Richa, A.W.: Broadcasting in unreliable radio networks. In: Proceedings of the 29th Symposium on Principles of Distributed Computing (PODC) (2010)
30. Kuhn, F., Lynch, N., Oshman, R.: Distributed computation in dynamic networks. In: Proceeedings of the 42nd Symposium on Theory of Computing (STOC) (2010)
31. Kuhn, F., Oshman, R.: Dynamic networks: models and algorithms. ACM SIGACT News **42**(1), 82–96 (2011)
32. Lynch, N., Newport, C.: A (truly) local broadcast layer for unreliable radio networks. In: Proceedings of the 34th Symposium on Principles of Distributed Computing (PODC) (2015)
33. Moscibroda, T., Wattenhofer, R.: The complexity of connectivity in wireless networks. In: Proceedings of the 25th Conference on Computer Communications (INFOCOM) (2006)
34. Newport, C.: Radio network lower bounds made easy. In: Proceedings of the 28th International Symposium on Distributed Computing (DISC) (2014)
35. Newport, C., Kotz, D., Yuan, Y., Gray, R.S., Liu, J., Elliott, C.: Experimental evaluation of wireless simulation assumptions. Simulation (2007)
36. O'Dell, R., Wattenhofer, R.: Information dissemination in highly dynamic graphs. In: Proceedings of Workshop on Foundations of Mobile Computing (DIALM-POMC) (2005)
37. Ramachandran, K., Sheriff, I., Belding, E., Almeroth, K.: Routing stability in static wireless mesh networks. In: Proceedings of the 8th International Conference on Passive and Active Network Measurement (2007)
38. Srinivasan, K., Kazandjieva, M.A., Agarwal, S., Levis, P.: The β-factor: measuring wireless link burstiness. In: Proceedings of the 6th Conference on Embedded Networked Sensor System (2008)
39. Yarvis, M.D., Conner, S.W., Krishnamurthy, L., Chhabra, J., Elliott, B., Mainwaring, A.: Real-world experiences with an interactive ad hoc sensor network. In: Proceedings of the Conference on Parallel Processing (2002)

Global Synchronization and Consensus Using Beeps in a Fault-Prone MAC

Kokouvi Hounkanli[1], Avery Miller[2(✉)], and Andrzej Pelc[1]

[1] Université du Québec en Outaouais, Gatineau, Canada
[2] University of Manitoba, Winnipeg, Canada
avery@averymiller.ca

Abstract. Global synchronization is an important prerequisite to many distributed tasks. Communication between processors proceeds in synchronous rounds. Processors are woken up in possibly different rounds. The clock of each processor starts in its wakeup round showing local round 0, and ticks once per round, incrementing the value of the local clock by one. The global round 0, unknown to processors, is the wakeup round of the earliest processor. Global synchronization (or establishing a global clock) means that each processor chooses a local clock round such that their chosen rounds all correspond to the same global round t.

We study the task of global synchronization in a Multiple Access Channel (MAC) prone to faults, under a very weak communication model called the *beeping model*. Some processors wake up spontaneously, in possibly different rounds decided by an adversary. In each round, an awake processor can either listen, i.e., stay silent, or beep, i.e., emit a signal. In each round, a fault can occur in the channel independently with constant probability $0 < p < 1$. In a fault-free round, an awake processor hears a beep if it listens in this round and if one or more other processors beep in this round. A processor still dormant in a fault-free round in which some other processor beeps is woken up by this beep and hears it. In a faulty round nothing is heard, regardless of the behaviour of the processors. An algorithm working with error probability at most ϵ, for a given $\epsilon > 0$, is called ϵ-*safe*. Our main result is the design and analysis, for any constant $\epsilon > 0$, of a deterministic ϵ-safe global synchronization algorithm that works in constant time in any fault-prone MAC using beeps.

As an application, we solve the consensus problem in a fault-prone MAC using beeps. Processors have input values from some set V and they have to decide the same value from this set. If all processors have the same input value, then they must all decide this value. Using global synchronization, we give a deterministic ϵ-safe consensus algorithm that works in time $O(\log w)$ in a fault-prone MAC, where w is the smallest input value of all participating processors. We show that this time cannot be improved, even when the MAC is fault-free.

Keywords: Global synchronization · Consensus · Multiple access channel · Fault · Beep

Partially supported by NSERC discovery grant 8136 – 2013 and by the Research Chair in Distributed Computing at the Université du Québec en Outaouais.

© Springer International Publishing AG 2017
M. Chrobak et al. (Eds.): ALGOSENSORS 2016, LNCS 10050, pp. 16–28, 2017.
DOI: 10.1007/978-3-319-53058-1_2

1   Introduction

1.1   The Problem

Global synchronization is an important prerequisite to many distributed tasks. Communication between processors proceeds in synchronous rounds. Processors are woken up in possibly different rounds. The clock of each processor starts in its wakeup round showing local round 0, and ticks once per round, incrementing the value of the local clock by one. The global round 0, unknown to processors, is the wakeup round of the earliest processor. Global synchronization (or establishing a global clock) means that each processor chooses a local clock round such that their chosen rounds all correspond to the same global round t. Achieving global synchronization permits to assume that in a subsequent task all processors start in the same round. This assumption is often used in solving various distributed problems. Global synchronization was assumed, e.g., in [10] for broadcasting and gossiping, in [14] for leader election, in [30] for minimum connected dominating set construction, and in [20] for the conflict resolution and the membership problem.

1.2   Model Description

We study the task of global synchronization in a fault-prone Multiple Access Channel (MAC): all processors can communicate directly, i.e., the underlying communication graph is complete. We adopt a very weak communication model called the *beeping model*. We assume that processors are fault free, while the MAC is prone to random faults. Faults in the channel may be due to some random noise occurring in the background. We assume that processors in the channel do not have access to any random generator. Some processors wake up spontaneously, in possibly different rounds decided by an adversary. In each round, an awake processor can either *listen*, i.e., stay silent, or *beep*, i.e., emit a signal. In each round, a fault can occur in the channel independently with constant probability $0 < p < 1$. The value of p is known by all processors. In a fault-free round, an awake processor hears a beep if it listens in this round and if one or more other processors beep in this round. In a faulty round, nothing is heard regardless of the behaviour of the processors. A processor that is still dormant in a fault-free round in which some other processor beeps is woken up by this beep and hears it.

The beeping model was introduced in [9] for vertex coloring, used in [1] to solve the MIS problem, and later used in [14,16] to solve leader election. The beeping model is widely applicable, as it makes small demands on communicating devices by relying only on carrier sensing. In fact, as mentioned in [9], beeps are an even weaker way of communicating than using one-bit messages: one-bit messages allow three different states (0,1 and no message), while beeps permit to differentiate only between a signal and its absence.

It has been noted that communication by beeps is a good model for various kinds of biological networks that arise in nature [24]. Given this motivation, it is especially important to understand how distributed coordination and communication can be carried out in a robust way to deal with unpredictable interference from the environment. We study deterministic global synchronization algorithms working in a probabilistic fault-prone MAC, which work with error probability at most ϵ, for a given $\epsilon > 0$. Such algorithms are called ϵ-safe. We assume that all processors know the value of ϵ.

1.3   Application

As an application of our global synchronization algorithm we solve the consensus problem in a fault-prone MAC, using beeps. Consensus is one of the fundamental tasks studied in distributed computing [21]. Processors have input values from some set V, and they have to decide the same value from this set. If all processors have the same input value, then they must all decide this value. Consensus has mostly been studied in the context of fault-tolerance. Either the communication between processors is assumed prone to faults [18,27,28], or processors themselves can be subject to crash [8,23] or Byzantine [25] faults. In the present paper, we study a scenario falling under the first of these variants.

We study the task of consensus defined precisely as follows [21]. Processors have input values from some set V of non-negative integers. The goal for all processors is to satisfy the following requirements.

Termination: all processors must output some value from V.

Agreement: all outputs must be equal.

Validity: if all input values are equal to v, then all output values must be equal to v.[1]

1.4   Our Results

Our main result is the design and analysis, for any constant $\epsilon > 0$, of a deterministic ϵ-safe global synchronization algorithm that works in constant time in any fault-prone MAC using beeps.

As an application we solve the consensus problem in a fault-prone MAC using beeps. Using global synchronization, we give a deterministic ϵ-safe consensus algorithm that works in time $O(\log w)$ in a fault-prone MAC, where w is the smallest input value of all participating processors. We show that this time cannot be improved, even when the MAC is fault-free. Moreover, we show how to reach consensus in the same round. Hence, as formulated in [23], we reach "double agreement, one on the decided value (data agreement) and one on the decision round (time agreement)".

Several proofs are omitted and will appear in the full version of the paper.

---

[1] Some authors use a stronger validity condition in which the output values must always be one of the input values, even if these are non-equal. In this paper we use the above formulation from [21].

1.5   Related Work

The Multiple Access Channel (MAC) is a popular and well-studied medium of communication. Most research concerning the MAC has been done under the radio communication model in which processors can send an entire message in a single round, and this message is heard by other processors if exactly one processor transmits, and all others listen in this round. This communication model is incomparable to the beeping model: on the one hand it is much stronger, as large messages (and not only beeps) can be sent in a single round, but on the other hand it is weaker, as it requires a unique transmitter in a round to make the transmission successful, while in the beeping model many beeps may be heard simultaneously. Broadcasting was studied in a MAC under the radio model, both in the deterministic [11,19] and in the randomized setting [5,29]. The throughput of a MAC under the radio model was studied in the situation where the channel could be jammed by an adversary, i.e., a collision is caused on the channel by the adversary at arbitrary times [2]. In [4], the authors give a randomized protocol for a MAC under the radio model that achieves constant throughput in the presence of an adversary that can arbitrarily jam the channel in a $(1 - \epsilon)$-fraction of the time slots.

The differences between local and global clocks for the wake-up problem were first studied in [15] and then in [6,9,12]. The communication model used in these papers was that of radio networks in which the main challenge is the occurrence of collisions between simultaneously received messages. Global synchronization is often used in the study of broadcasting in radio networks (cf. [12]). The previously cited papers [10,14,30] used the assumption of global synchronization in the beeping model in multi-hop networks. In [22] the authors compared different variations of the beeping model, also assuming global synchronization. The authors of [20] used the assumption of global synchronization in the beeping model in a multiple access channel. All these papers assumed that communication is fault free.

Consensus is a classic problem in distributed computing, mostly studied assuming that processors communicate by shared variables or through message passing networks [3,21]. See the recent book [26] for a comprehensive survey of the literature on consensus, mostly concerning processor faults. In [17], the authors showed a randomized consensus for crash faults with optimal communication complexity. In [8], the feasibility and complexity of consensus in a multiple access channel (MAC) with simultaneous wake-up and crash failures were studied in the context of different collision detectors. Consensus (without faults) in a MAC with different wake-up times was studied in [13]. The authors also investigated the impact of a global clock on the time efficiency of consensus. Consensus in the quantum setting has been studied, e.g., in [7]. To the best of our knowledge, neither global synchronization nor consensus with faulty beeps have ever been studied before.

2   Global Synchronization

In this section, we provide an algorithm GlobalSync that establishes a global clock. Upon its wake-up, each processor in the channel executes GlobalSync with its local clock initialized to 0. The round in which the first wake-up occurs is defined as global round 0. Processors are not aware of the relationship between their local clock values and this global round. Establishing a global clock means that all processors in the channel exit GlobalSync in the same global round.

Fix any constant $\epsilon > 0$. Let γ be a constant such that $p^\gamma < \frac{\epsilon}{4}$. Hence, in a sequence of γ consecutive rounds of beeps, at least one of these beeps occurs in a fault-free round with probability at least $1 - \frac{\epsilon}{4}$.

We describe Algorithm GlobalSync whose aim is to ensure that all processors agree on a common global round, i.e. they establish a global clock. At a high level, the algorithm proceeds as follows. A processor that wakes up spontaneously beeps periodically trying to wake up all other processors that are still dormant. These beeps will be called *alarm beeps*. They are separated by time intervals of increasing size, which prevents an adversary from setting wake-up times so that all alarm beeps are aligned. In the intervals between alarm beeps, the processor is waiting for a response from other processors to indicate that they heard an alarm beep. If a large enough number of such intervals occur without any response, then the processor assumes that the entire channel was woken up at the same time, and a global round is chosen as the round in which the next alarm beep is scheduled. Otherwise, if a beep was heard in one of these intervals, the processor listens for 2γ consecutive rounds and then beeps for 2γ consecutive rounds. Similarly, a processor woken up by a beep listens for 2γ consecutive rounds and then beeps for 2γ consecutive rounds. The global round chosen by the algorithm is the round $r + 4\gamma + 1$, where r is the first round when an alarm beep was heard by some processor. The difficulty is for each processor to determine the round r. This is because, when a beep is heard, there are two possible cases: such a beep may be an alarm beep from another processor, or may be in response to an alarm beep. We overcome this difficulty as follows. Time is divided into blocks of 2γ consecutive rounds. If a single beep is heard in a block, the processor concludes that it was an alarm beep; if more than one beep is heard in a block, the processor concludes that these beeps were in response to an alarm beep. We will prove that such conclusions are correct with sufficiently high probability. Finally, each processor considers the first round s in which it heard a beep. If this beep was an alarm beep, the processor sets $r = s$. If this beep was in response to an alarm beep, the processor sets r to be the most recent round before s in which it beeped.

We now provide the details of Algorithm GlobalSync. The following procedure provides an aggregate count of the beeps recently heard by a processor. More specifically, for a given round t', the next 4γ rounds are treated as two blocks of 2γ rounds each, and for each block, the cases of 0, 1, or more beeps are distinguished.

Algorithm 1. listenVector(t')

1: $h_1 \leftarrow 0$
2: $h_2 \leftarrow 0$
3: $num_1 \leftarrow$ number of beeps heard in rounds $t', t' + 1, \ldots, t' + 2\gamma - 1$
4: $num_2 \leftarrow$ number of beeps heard in rounds $t' + 2\gamma, \ldots, t' + 4\gamma - 1$
5: **if** $num_1 = 1$, **then** $h_1 \leftarrow 1$
6: **if** $num_1 > 1$, **then** $h_1 \leftarrow *$
7: **if** $num_2 = 1$, **then** $h_2 \leftarrow 1$
8: **if** $num_2 > 1$, **then** $h_2 \leftarrow *$
9: **return** $[h_1\ h_2]$

Below we give the pseudocode of Algorithm GlobalSync using the above procedure.

Algorithm 2. GlobalSync

1: **if** woken up by a beep in some round *heard*:   ▷ woken up by beep
2:     beep 2γ consecutive rounds starting at round $heard + 2\gamma + 1$
3:     $syncRound \leftarrow heard + 4\gamma + 1$
4: **else**:   ▷ woken up spontaneously
5:     $i \leftarrow 0$
6:     $myNextBeep \leftarrow 0$
7:     **repeat**:
8:         $myCurrentBeep \leftarrow myNextBeep$
9:         beep in round $myCurrentBeep$
10:        $i \leftarrow i + 1$
11:        $myNextBeep \leftarrow myCurrentBeep + 4\gamma + i$
12:     **until** $(i = 3\gamma)$ or (a beep is heard in one of $\{myCurrentBeep + 1, \ldots, myNextBeep - 1\}$)
13:     **if** $i = 3\gamma$:
14:         $syncRound \leftarrow myNextBeep$
15:     **else**:
16:         $heard \leftarrow$ first round after $myCurrentBeep$ in which a beep was heard
17:         $[h_1\ h_2] = listenVector(myCurrentBeep + 1)$
18:         **if** $[h_1\ h_2] \in \{[0\ 0], [0\ 1], [1\ 0], [1\ 1], [1\ *]\}$:
19:             beep 2γ consecutive rounds starting at round $heard + 2\gamma + 1$
20:             $syncRound \leftarrow heard + 4\gamma + 1$
21:         **if** $[h_1\ h_2] \in \{[0\ *], [*\ 0], [*\ 1], [*\ *]\}$:
22:             $syncRound \leftarrow myCurrentBeep + 4\gamma + 1$
23: wait until round $syncRound$ and **exit**

In the analysis of Algorithm GlobalSync we refer to global rounds, but it should be recalled that processors in the channel do not have access to the global clock values: all a processor sees is its local clock. The following fact follows from the algorithm description by induction on i.

Fact 1. *At the end of each loop iteration i, the variable myNextBeep is equal to $4\gamma i + \sum_{k=1}^{i} k$. Further, if a processor is woken up at time t, then, at the end of each loop iteration i, myNextBeep is equal to the local clock value corresponding to the global round $t + 4\gamma i + \sum_{k=1}^{i} k$.*

We say that a processor is *lonely in round t* if it has not heard a beep in any round up to and including round t. Using Fact 1, we can determine the number of rounds that elapse before a lonely processor beeps a given number of times.

Fact 2. *Suppose that a processor v wakes up spontaneously in round t_1. If v is lonely in round $t_1 + 4\gamma i + i(i+1)/2$, then v has beeped exactly i times before this round.*

In order to prove the correctness of the algorithm, we first consider the case when all processors wake up spontaneously in the same round.

Lemma 1. *Suppose that all processors wake up spontaneously in the same global round t_1. With probability 1, all processors terminate their execution of* GlobalSync *in global round $t_1 + 12\gamma^2 + (3\gamma)(3\gamma + 1)/2$.*

Proof. Since every processor is woken up spontaneously in global round t_1, the **if** condition on line 1 evaluates to false at every processor. Therefore, all processors execute the loop at line 7. In particular, this means that all processors beep in their local round 0. By Fact 1, at the end of each loop iteration, the variable *myNextBeep* at every processor is equal to the local clock value corresponding to the global round $t_1 + 4\gamma i + \sum_{k=1}^{i} k$. It follows that all processors beep in the same rounds. In particular, this means that no processor ever hears a beep. Thus, at every processor, the loop exits with $i = 3\gamma$. So, the **if** condition on line 13 evaluates to true. By line 14, each processor sets *syncRound* to the value $4\gamma(3\gamma) + \sum_{k=1}^{3\gamma} k = 12\gamma^2 + (3\gamma)(3\gamma + 1)/2$, which is their local clock value that corresponds to the global round $t_1 + 12\gamma^2 + (3\gamma)(3\gamma + 1)/2$. Therefore, all processors terminate their execution of GlobalSync in global round $t_1 + 12\gamma^2 + (3\gamma)(3\gamma + 1)/2$. $\square$

Note that, when our algorithm is executed in the case where all processors wake up spontaneously in the same round, no processor ever hears a beep, and, after a fixed length of silence, all processors terminate their execution of GlobalSync. In the case where not all processors wake up spontaneously in the same round, if the same fixed length of silence is observed by all processors, then, again, all processors will terminate their execution of GlobalSync, but this time in different rounds. This would be a bad case for our algorithm. We now show that, with sufficiently high probability, such a bad case does not occur, i.e., that there exists some round t^* in which a beep is heard by some processor.

Lemma 2. *Suppose that not all processors wake up spontaneously in the same round, and suppose that the first spontaneous wake-up occurs in some round t_1. With probability at least $(1 - \frac{\epsilon}{2})$, there exists a global round $t^* \leq t_1 + 12\gamma^2 + (3\gamma)(3\gamma + 1)/2$ in which all of the following hold: no processor has terminated its execution of* GlobalSync*, at least one processor beeps, at least one processor listens, and no fault occurs.*

We now proceed to prove the correctness of our algorithm for the case where not all processors wake up spontaneously in the same round. We will be able to do so when there exists a global round t^* satisfying the conditions specified in Lemma 2. The next lemma shows that all processors terminate their execution of GlobalSync in the same global round soon after t^*.

Lemma 3. *Suppose that not all processors wake up spontaneously in the same round. Let t^* be the first global round in which all of the following hold: no processor has terminated its execution of GlobalSync, at least one processor beeps, at least one processor listens, and no fault occurs. With probability at least $(1 - \frac{\epsilon}{2})$, all processors terminate their execution of GlobalSync in global round $t^* + 4\gamma + 1$.*

Finally, we show that Algorithm GlobalSync runs in constant time and fails with probability at most ϵ for any given constant $\epsilon > 0$.

Theorem 3. *Fix any constant $\epsilon > 0$. With probability at least $1 - \epsilon$, all processors terminate Algorithm GlobalSync in the same global round sync, which occurs $O(1)$ rounds after the first wake-up.*

Proof. Let t_1 be the first round in which a wake-up occurs. In the case where all processors wake up spontaneously in the same round, Lemma 1 implies that, with probability 1, all processors terminate Algorithm GlobalSync in global round sync $= t_1 + 12\gamma^2 + (3\gamma)(3\gamma + 1)/2$. In the case where not all processors wake up spontaneously in the same round, Lemmas 2 and 3 imply that all processors terminate Algorithm GlobalSync in global round sync $= t^* + 4\gamma + 1$, where $t^* \leq t_1 + 12\gamma^2 + (3\gamma)(3\gamma + 1)/2$, with error probability at most $\frac{\epsilon}{2} + \frac{\epsilon}{2} = \epsilon$. □

3   Consensus

In this section, we provide a deterministic decision procedure that achieves consensus assuming that global synchronization has been done previously. It is performed after Algorithm GlobalSync and has the following property. Let sync be the global round in which all processors in the channel terminate their execution of Algorithm GlobalSync. Algorithm Decision achieves consensus with error probability at most ϵ in the global round $s =$ sync $+ O(\log w)$, where w is the smallest of all input values of processors in the channel.

Consider the input value val of a processor v and let $\mu = (a_1, \ldots, a_m)$ be its binary representation. We transform the sequence μ by replacing each bit 1 by (10), each bit 0 by (01) and appending (11) at the end. Hence the transformed sequence is $(c_1, \ldots, c_{2m+2})$, where
$c_i = 1$, for $i \in \{2m + 1, 2m + 2\}$, and,
for $j = 1, \ldots, m$:
$c_{2j-1} = 1$ and $c_{2j} = 0$, if $a_j = 1$,
$c_{2j-1} = 0$ and $c_{2j} = 1$, if $a_j = 0$.

The sequence $(c_1, \ldots, c_{2m+2})$ is called the *transformed input value* of processor v and is denoted by val^*. Notice that if the input values of two processors are different, then there exists an index for which the corresponding bits of their transformed input values differ (this is not necessarily the case for the original input values, since one of the binary representations might be a prefix of the other).

The high-level idea of Algorithm Decision is the following. A processor beeps and listens in time intervals of prescribed length, starting in global round $\mathtt{sync} + 1$, according to its transformed input value. If it does not hear any beep, it concludes that all input values are identical and outputs its input value. Otherwise, it concludes that there are different input values and then outputs a default value. We will prove that these conclusions are correct with probability at least $1 - \epsilon$, and that all processors make the decision in a common global round $s = \mathtt{sync} + O(\log w)$.

We now give the pseudocode of the algorithm executed by a processor whose input value is val. We assume that the algorithm is started in global round $\mathtt{sync} + 1$, and we let r be the processor's local clock value corresponding to the global round $\mathtt{sync}$. Let x be the smallest positive integer such that $p^x < \epsilon/2$. Let val_0 be the smallest integer in V, which we will use as the default decision value.

Algorithm 3. Decision

1: $(c_1, \ldots, c_k) \leftarrow val^*$
2: $i \leftarrow 1$
3: $heard \leftarrow false$
4: **while** ($heard = false$ **and** $i \leq k$) **do**
5:     **if** $c_i = 1$ **then**
6:         beep for x rounds and then listen for x rounds
7:     **if** $c_i = 0$ **then**
8:         listen for x rounds and then beep for x rounds
9:     **if** a beep was heard **then** $heard \leftarrow true$
10:     $i \leftarrow i + 1$
11: **if** $heard = false$ **then** output val in round $r + 2(i-1)x + 1$
12: **else** output val_0 in round $r + 2(i-1)x + 1$

The following result shows that, with error probability at most ϵ, upon completion of Algorithm Decision, all processors in the channel correctly solve consensus in the same round, and this round occurs $O(\log w)$ rounds after global round $\mathtt{sync}$, where w is the smallest of all input values of processors in the channel.

Theorem 4. *Let* $\mathtt{sync}$ *be the common global round in which all processors terminate their execution of Algorithm* GlobalSync, *and let* w *be the smallest of all input values of processors in the channel. There exists a global round* $s = \mathtt{sync} + O(\log w)$ *such that, with probability at least* $1 - \epsilon$, *upon completion of Algorithm* Decision, *all processors in the channel output the same value in global round* s, *and this value is their common input value if all input values were identical.*

Proof. First, suppose that the input values of all processors in the channel are identical. Let $k \in O(\log w)$ be the length of their common transformed input value. Then each processor leaves the **while** loop with the value of the variable *heard* equal to false, and consequently, at line 11, it outputs the common input value in round $r + 2xk + 1$, which is its local clock value corresponding to global round $\texttt{sync} + 2xk + 1$. Since x is a constant, we have $2xk + 1 \in O(\log w)$, which concludes the proof in this case.

In the remainder of the proof, we suppose that there are at least two distinct input values. Let k_1 be the length of the transformed input value w^* corresponding to the input value w. Consider all transformed input values of processors in the channel, and let $j \leq k_1$ be the first index in which two of these transformed input values differ.

For any $t > 0$, let A_t be the global time interval $\{\texttt{sync} + 2x(t - 1) + 1, \ldots, \texttt{sync} + 2x(t - 1) + x\}$, and let B_t be the global time interval $\{\texttt{sync} + 2x(t - 1) + x + 1, \ldots, \texttt{sync} + 2x(t - 1) + 2x\}$. Let E be the event that at least one round in the time interval A_j is fault free and at least one round in the time interval B_j is fault free. By the definition of x, the probability of event E is at least $1 - \epsilon$. Suppose that event E holds. By the choice of j, no beep was heard in the channel in global rounds $\{\texttt{sync} + 1, \ldots, \texttt{sync} + 2x(j - 1)\}$, hence all processors in the channel participate in the jth iteration of the loop. Consider any processor v for which the jth bit of its transformed input value is 0 and any processor v' for which the jth bit of its transformed input value is 1. Processor v listens in all rounds of A_j and beeps in all rounds of B_j, whereas processor v' beeps in all rounds of A_j and listens in all rounds of B_j. Hence, v hears at least one beep in the time interval A_j, and v' hears at least one beep in the time interval B_j. Therefore, both v and v' set *heard* equal to true in iteration j of the **while** loop. Consequently, each processor outputs the default value val_0 at line 12 in round $r + 2xj + 1$, which is its local clock value corresponding to global round $\texttt{sync} + 2xj + 1$. Since x is constant and $j \leq k_1 \in O(\log w)$, we have $2xj + 1 \in O(\log w)$, which concludes the proof in the case where there are at least two distinct input values. $\square$

Finally, given a bound $\epsilon > 0$ on error probability of consensus, we first run Algorithm GlobalSync and then Algorithm Decision, each with error probability bound $\frac{\epsilon}{2}$, to get the following corollary.

Corollary 1. *Fix any constant $\epsilon > 0$. With error probability at most ϵ, consensus can be solved deterministically using beeps in a fault-prone MAC in time $O(\log w)$, where w is the smallest of all input values of processors in the channel.*

We conclude this section by showing that, even in a model where every round in the MAC is fault-free and all processors are woken up spontaneously in the same round, deterministic consensus with m-bit inputs requires $\Omega(m)$ rounds, which implies that our consensus algorithm has optimal time complexity.

Theorem 5. *Consensus with m-bit inputs in a fault-free MAC with beeps requires $\Omega(m)$ rounds.*

Proof. Consider any consensus algorithm $\mathcal{A}$. Assume that, for every m-bit input value s, the execution of $\mathcal{A}$ with input s by a single processor on the channel uses $o(m)$ rounds. For any input s, let Pattern(s) be the beeping pattern of a processor that is alone on the channel and executes $\mathcal{A}$ with input s. By the Pigeonhole Principle, there exist distinct m-bit inputs a and b such that Pattern(a) = Pattern(b).

For each $s \in \{a, b\}$, let α_s be the execution of $\mathcal{A}$ in the case where a processor v_s is alone on the channel and is given input s. By Validity, for each $s \in \{a, b\}$, at the end of execution α_s, processor v_s must output s. Next, consider the execution $\alpha_{a,b}$ of $\mathcal{A}$ in the case where processors v_a and v_b are on the channel and are given inputs a and b, respectively. Since Pattern(a) = Pattern(b), it follows that executions α_a and $\alpha_{a,b}$ are indistinguishable to processor v_a, and that executions α_b and $\alpha_{a,b}$ are indistinguishable to processor v_b. Therefore, in execution $\alpha_{a,b}$, processor v_a outputs a and processor v_b outputs b, which contradicts Agreement. Therefore, we incorrectly assumed that, for every m-bit input s, the execution of $\mathcal{A}$ with input s by a single processor on the channel uses $o(m)$ rounds. It follows that there exists an execution of $\mathcal{A}$ that uses $\Omega(m)$ rounds, as claimed.   □

4   Open Questions

Our work leaves open several interesting directions for future research. One question is how to achieve global synchronization in multi-hop networks. One could also ask how to design an efficient synchronization algorithm if the probability p of channel failures is not known by the processors. Probably the most intriguing direction is to think about different kinds of faults. Rather than jamming faults that affect the entire channel, we can think about faults that occur at individual processors, e.g., if a processor's beep fails and does not get transmitted on the channel, or if a listening processor fails to hear a beep that is transmitted by the channel.

References

1. Afek, Y., Alon, N., Bar-Joseph, Z., Cornejo, A., Haeupler, B., Kuhn, F.: Beeping a maximal independent set. In: Peleg, D. (ed.) DISC 2011. LNCS, vol. 6950, pp. 32–50. Springer, Heidelberg (2011). doi:10.1007/978-3-642-24100-0_3
2. Anantharamu, L., Chlebus, B.S., Kowalski, D.R., Rokicki, M.A.: Medium access control for adversarial channels with jamming. In: Kosowski, A., Yamashita, M. (eds.) SIROCCO 2011. LNCS, vol. 6796, pp. 89–100. Springer, Heidelberg (2011). doi:10.1007/978-3-642-22212-2_9
3. Attiya, H., Welch, J.: Distributed Computing. John Wiley and Sons Inc., Chichester (2004)
4. Awerbuch, B., Richa, A.W., Scheideler, C., Schmid, S., Zhang, J.: Principles of robust medium access and an application to leader election. ACM Trans. Algorithms **10**(4), 24:1–24:26 (2014)
5. Bar-Yehuda, R., Goldreich, O., Itai, A.: On the time complexity of broadcast in radio networks: an exponential gap between determinism and randomization. J. Comput. Syst. Sci. **45**, 104–126 (1992)

6. Chlebus, B.S., Gąsieniec, L., Kowalski, D.R., Radzik, T.: On the wake-up problem in radio networks. In: Caires, L., Italiano, G.F., Monteiro, L., Palamidessi, C., Yung, M. (eds.) ICALP 2005. LNCS, vol. 3580, pp. 347–359. Springer, Heidelberg (2005). doi:10.1007/11523468_29

7. Chlebus, B.S., Kowalski, D.R., Strojnowski, M.: Scalable quantum consensus for crash failures. In: Lynch, N.A., Shvartsman, A.A. (eds.) DISC 2010. LNCS, vol. 6343, pp. 236–250. Springer, Heidelberg (2010). doi:10.1007/978-3-642-15763-9_24

8. Chockler, G., Demirbas, M., Gilbert, S., Lynch, N.A., Newport, C.C., Nolte, T.: Consensus and collision detectors in radio networks. Distrib. Comput. **21**, 55–84 (2008)

9. Cornejo, A., Kuhn, F.: Deploying wireless networks with beeps. In: Lynch, N.A., Shvartsman, A.A. (eds.) DISC 2010. LNCS, vol. 6343, pp. 148–162. Springer, Heidelberg (2010). doi:10.1007/978-3-642-15763-9_15

10. Czumaj, A., Davis, P.: Communicating with beeps, arXiv:1505.06107 [cs.DC] (2015)

11. Clementi, A.E.F., Monti, A., Silvestri, R.: Selective families, superimposed codes, and broadcasting on unknown radio networks. In: Proceedings of the 12th Annual ACM-SIAM Symposium on Discrete Algorithms (SODA 2001), pp. 709–718 (2001)

12. Czumaj, A., Rytter, W.: Broadcasting algorithms in radio networks with unknown topology. In: Proceedings of the 44th IEEE Symposium on Foundations of Computer Science (FOCS 2003), pp. 492–501 (2003)

13. Czyzowicz, J., Gąsieniec, L., Kowalski, D.R., Pelc, A.: Consensus and mutual exclusion in a multiple access channel. In: Keidar, I. (ed.) DISC 2009. LNCS, vol. 5805, pp. 512–526. Springer, Heidelberg (2009). doi:10.1007/978-3-642-04355-0_51

14. Förster, K.-T., Seidel, J., Wattenhofer, R.: Deterministic leader election in multi-hop beeping networks. In: Kuhn, F. (ed.) DISC 2014. LNCS, vol. 8784, pp. 212–226. Springer, Heidelberg (2014). doi:10.1007/978-3-662-45174-8_15

15. Gasieniec, L., Pelc, A., Peleg, D.: The wakeup problem in synchronous broadcast systems. SIAM J. Discrete Math. **14**, 207–222 (2001)

16. Ghaffari, M., Haeupler, B.: Near optimal leader election in multi-hop radio networks. In: Proceedings of the 24th Annual ACM-SIAM Symposium on Discrete Algorithms (SODA 2013), pp. 748–766 (2013)

17. Gilbert, S., Kowalski, D.R.: Distributed agreement with optimal communication complexity. In: Proceedings of the 21st ACM-SIAM Symposium on Discrete Algorithms (SODA 2010), pp. 965–977 (2010)

18. Gray, J.N.: Notes on data base operating systems. In: Bayer, R., Graham, R.M., Seegmüller, G. (eds.) Operating Systems an Advanced Course. LNCS, vol. 60, pp. 393–481. Springer, Heidelberg (1978). doi:10.1007/3-540-08755-9_9

19. Greenberg, A.G., Winograd, S.: A lower bound on the time needed in the worst case to resolve conflicts deterministically in multiple access channels. J. ACM **32**, 589–596 (1985)

20. Huang, B., Moscibroda, T.: Conflict resolution and membership problem in beeping channels. In: Afek, Y. (ed.) DISC 2013. LNCS, vol. 8205, pp. 314–328. Springer, Heidelberg (2013). doi:10.1007/978-3-642-41527-2_22

21. Lynch, N.A.: Distributed Algorithms. Morgan Kaufmann Publ. Inc., San Francisco (1996)

22. Métivier, Y., Robson, J.M., Zemmari, A.: On distributed computing with beeps, arXiv:1507.02721 [cs.DC] (2015)

23. Moses, Y., Raynal, M.: Revisiting simultaneous consensus with crash failures. J. Parallel Distrib. Comput. **69**, 400–409 (2009)

24. Navlakha, S., Bar-Joseph, Z.: Distributed information processing in biological and computational systems. Commun. ACM **58**(1), 94–102 (2014)
25. Pease, M.C., Shostak, R.E., Lamport, L.: Reaching agreement in the presence of faults. J. ACM **27**, 228–234 (1980)
26. Raynal, M.: Fault-Tolerant Agreement in Synchronous Distributed Systems. Morgan & Claypool Publishers, San Francisco (2010)
27. Santoro, N., Widmayer, P.: Time is not a healer. In: Monien, B., Cori, R. (eds.) STACS 1989. LNCS, vol. 349, pp. 304–313. Springer, Heidelberg (1989). doi:10.1007/BFb0028994
28. Santoro, N., Widmayer, P.: Distributed function evaluation in presence of transmission faults. In: Proceedings of the International Symposium on Algorithms (SIGAL 1990), pp. 358–367 (1990)
29. Willard, D.E.: Log-logarithmic selection resolution protocols in a multiple access channel. SIAM J. Comput. **15**, 468–477 (1986)
30. Yu, Y., Jia, L., Yu, D., Li, G., Cheng, X.: Minimum connected dominating set construction in wireless networks under the beeping model. In: Proceedings of the IEEE Conference on Computer Communications (INFOCOM 2015), pp. 972–980 (2015)

Vertex Coloring with Communication and Local Memory Constraints in Synchronous Broadcast Networks

Hicham Lakhlef[1], Michel Raynal[1,2($\boxtimes$)], and François Taïani[1]

[1] IRISA, Université de Rennes, Rennes, France
{hicham.lakhlef,raynal,francois.taiani}@irisa.fr
[2] Institut Universitaire de France, Paris, France

Abstract. This paper considers the broadcast/receive communication model in which message collisions and message conflicts can occur because processes share frequency bands. (A collision occurs when, during the same round, messages are sent to the same process by too many neighbors. A conflict occurs when a process and one of its neighbors broadcast during the same round.) More precisely, the paper considers the case where, during a round, a process may either broadcast a message to its neighbors or receive a message from at most m of them. This captures communication-related constraints or a local memory constraint stating that, whatever the number of neighbors of a process, its local memory allows it to receive and store at most m messages during each round. The paper defines first the corresponding generic vertex multi-coloring problem (a vertex can have several colors). It focuses then on tree networks, for which it presents a lower bound on the number of colors K that are necessary (namely, $K = \lceil \frac{\Delta}{m} \rceil + 1$, where Δ is the maximal degree of the communication graph), and an associated coloring algorithm, which is optimal with respect to K.

Keywords: Broadcast/receive · Bounded local memory · Collision-freedom · Distributed algorithm · Message-passing · Multi-coloring · Network traversal · Scalability · Synchronous system · Tree network · Vertex coloring

1 Introduction

Distributed message-passing synchronous systems. From a structural point of view, a message-passing system can be represented by a graph, whose vertices are the processes, and whose edges are the communication channels. It is naturally assumed that the graph is connected.

Differently from asynchronous systems, where there is no notion of global time accessible to the processes, synchronous message-passing systems are characterized by upper bounds on message transfer delays and processing times. Algorithms for such systems are usually designed according to the round-based programming paradigm. The processes execute a sequence of synchronous

© Springer International Publishing AG 2017
M. Chrobak et al. (Eds.): ALGOSENSORS 2016, LNCS 10050, pp. 29–44, 2017.
DOI: 10.1007/978-3-319-53058-1_3

rounds, such that, at every round, each process first sends a message to its neighbors, then receives messages from them, and finally executes a local computation, which depends on its local state and the messages it has received. The fundamental synchrony property of this model is that every message is received in the round in which it was sent. The progress from one round to the next is a built-in mechanism provided by the model. Algorithms suited to reliable synchronous systems can be found in several textbooks (e.g., [22,24])[1]. When considering reliable synchronous systems, an important issue is the notion of local algorithm. Those are the algorithms whose time complexity (measured by the number of rounds) is smaller than the graph diameter [1,19].

Distributed graph coloring in point-to-point synchronous systems. One of the most studied graph problems in the context of an n-process reliable synchronous system is the vertex coloring problem, namely any process must obtain a color, such that neighbor processes must have different colors (distance-1 coloring), and the total number of colors is reasonably "small". More generally, the distance-k coloring problem requires that no two processes at distance less or equal to k, have the same color. When considering sequential computing, the optimal distance-1 coloring problem is NP-complete [12].

When considering the distance-1 coloring problem in an n-process reliable synchronous system, it has been shown that, if the communication graph can be logically oriented such that each process has only one predecessor (e.g., a tree or a ring), $O(\log^* n)$ rounds are necessary and sufficient to color the processes with at most three colors [10,19][2]. Other distance-1 coloring algorithms are described in several articles (e.g. [3–5,14,17]). They differ in the number of rounds they need and in the number of colors they use to implement distance-1 coloring. Let Δ be the maximal degree of the graph (the degree of a vertex is the number of its neighbors). For instance [5] presents a vertex coloring algorithm which uses $(\Delta + 1)$ colors which requires $O(\Delta + \log^* n)$ rounds. An algorithm is described in [14] for trees, which uses three colors and $O(\log^* n)$ rounds. The algorithm presented in [17] requires $O(\Delta \log \Delta + \log^* n)$ rounds. These algorithms assume that the processes have distinct identities[3], which define their initial colors. They proceed iteratively, each round reducing the total number of colors. Distributed distance-2 and distance-3 coloring algorithms, suited to various synchronous models, are presented in [6–9,11,13,15].

Motivation and content of the paper. The previous reliable synchronous system model assumes that there is a dedicated (e.g., wired) bi-directional communication channel between each pair of neighbor processes. By contrast, this paper

---

[1] The case where processes may exhibit faulty behaviors (such as crashes or Byzantine failures) is addressed in several books (e.g., [2,20,22,23]).

[2] $\log^* n$ is the number of times the function log needs to be iteratively applied in $\log(\log(\log(...(\log n))))$ to obtain a value ≤ 2. As an example, if n is the number of atoms in the universe, $\log^* n \simeq 5$.

[3] Some initial asymmetry is necessary to solve *breaking symmetry problems* with a deterministic algorithm.

considers a broadcast/receive communication model in which there is no dedicated communication medium between each pair of neighbor processes. This covers practical system deployments, such as wireless networks and sensor networks. In such networks, the prevention of collisions (several neighbors of the same process broadcast during the same round), or conflicts (a process and one of its neighbors issue a broadcast during the same round), does not come for free. In particular, round-based algorithms that seek to provide deterministic communication guarantees in these systems must be collision and conflict-free (*C2*-free in short).

We are interested in this paper to offer a programming model in which, at any round, a process can either broadcast a message to its neighbors (conflict-freedom), or receive messages from at most m of its neighbors (m-collision-freedom). This means that we want to give users a round-based programming abstraction guaranteeing conflict-freedom and a weakened form of collision-freedom, that we encapsulate under the name C2m-freedom (if $m = 1$, we have basic C2-freedom).

The ability to simultaneously receive messages from multiple neighbors can be realized in practice by exploiting multiple frequency channels[4]. The parameter $m \geq 1$ is motivated by the following observations. While a process (e.g., a sensor) may have many neighbors, it can have constraints on the number of its reception channels, or constraints on its local memory, that, at each round, allow it to receive and store messages from only a bounded subset of its neighbors, namely m of them ($m = 1$, gives the classic C2-free model, while $m \geq \Delta$ assumes no collision can occur as in the classic broadcast/receive model presented previously). This "bounded memory" system parameter can be seen as a scalability parameter, which allows the degree of a process (number of its neighbors) to be decoupled from its local memory size.

C2m-freedom can be easily translated as a coloring problem, where any two neighbors must have different colors (conflict-freedom), and any process has at most m neighbors with the same color (m-collision-freedom). Once such a coloring is realized, message consistency is ensured by restricting the processes to broadcast messages only at the rounds associated with their color. While it is correct, such a solution can be improved, to allow more communication to occur during each round. More precisely, while guaranteeing C2m-freedom, it is possible to allow processes to broadcast at additional rounds, by allocating multiple colors to processes. From a graph coloring point of view, this means that, instead of only one color, a set of colors can be associated with each process, while obeying the following two constraints: (a) for any two neighbor processes, the intersection of their color sets must remain empty; and (b) given any process, no color must appear in the color sets of more than m of its neighbors.

---

[4] Depending on the underlying hardware (e.g., multi-frequency bandwidth, duplexer, diplexer), variants of this broadcast/receive communication pattern can be envisaged. The algorithms presented in this paper can be modified to take them into account.

We call *Coloring with Communication/Memory Constraints* (CCMC) the coloring problem described above. This problem is denoted CCMC $(n, m, K, \geq 1)$, where n is the number of processes (vertices), m is the bound on each local memory (bound on the number of simultaneous communication from a reception point of view), and K the maximal number of colors that are allowed. "≥ 1" means that there is no constraint on the number of colors that can be assigned to a process. From a technical point of view, the paper focuses on tree networks. It presents a lower bound on the value of K for these communication graphs, and an algorithm, optimal with respect to K, which solves both instances of CCMC.

CCMC is closely related to, but different from the β-frugal coloring problem, first introduced in [16]. β-frugal coloring considers a traditional distance-1 coloring in which a color is used at most β times in each node's neighborhood. A first difference lies in the number of colors assigned to each node. β-frugal coloring is a special case of CCMC, in which nodes are assigned a single color, rather than a set ($CCMC(n, \beta, K, 1)$ with our notation). A second, more fundamental difference lies in the computing model we adopt in our work, which assumes (undetected) conflicts and collisions between nodes. These conflicts and collisions complicate the task of individual nodes that produce a coloring while coordinating their communications to avoid message losses. Finally, while existing works on β-frugal coloring [16,21] have primarily focused on asymptotic bounds for graphs exhibiting sufficiently large values of Δ (e.g. $\Delta > \Delta_0 = e^{10^7} \approx 10^{4342944}$ [16]), we do not impose any such minimal bound on Δ in our work, providing results that apply to graphs actually encountered in practical systems.

Roadmap. The paper is made up of Sect. 7 sections. Section 2 presents the underlying system model. Section 3 formally defines the CCMC problem. Then, considering tree networks, whose roots are dynamically defined, Sect. 4 presents a lower bound on K for CCMC$(n, m, K, 1)$ and CCMC$(n, m, K, \geq 1)$ to be solved. Section 5 presents then a K-optimal algorithm solving CCMC$(n, m, K, \geq 1)$ (from which a solution to CCMC$(n, m, K, 1)$ can be easily obtained). Sect. 6 presents a proof of the algorithm (missing proofs can be found in [18]). Finally, Sect. 7 concludes the paper.

2   Synchronous Broadcast/Receive Model

Processes, initial knowledge, and the communication graph. The system model consists of n sequential processes denoted $p_1, ..., p_n$, connected by a connected communication graph. When considering a process p_i, $1 \leq i \leq n$, the integer i is called its index. Indexes are not known by the processes. They are only a notation convenience used to distinguish processes and their local variables.

Each process p_i has an identity id_i, which is known only by itself and its neighbors (processes at distance 1 from it). The constant $neighbors_i$ is a local set, known only by p_i, including the identities of its neighbors (and only them). In order for a process p_i not to confuse its neighbors, it is assumed that no two processes at distance less than or equal to 2 have the same identity. Hence, any two processes at distance greater than 2 can have the very same identity.

Δ_i denotes the degree of process p_i (i.e. $|neighbors_i|$) and Δ denotes the maximal degree of the graph ($\max\{\Delta_1, \cdots, \Delta_n\}$). While each process p_i knows Δ_i, no process knows Δ (a process p_x such that $\Delta_x = \Delta$ does not know that Δ_x is Δ).

Timing model. Processing durations are assumed equal to 0. This is justified by the following observations: (a) the duration of local computations is negligible with respect to message transfer delays, and (b) the processing duration of a message may be considered as a part of its transfer delay. Communication is synchronous in the sense that there is an upper bound D on message transfer delays, and this bound is known by all the processes (global knowledge). From an algorithm design point of view, we consider that there is a global clock, denoted $CLOCK$, which is increased by 1, after each period of D physical time units. Each value of $CLOCK$ defines what is usually called a *time slot* or a *round*.

Communication operations. The processes are provided with two operations denoted bcast() and receive(). A process p_i invokes bcast TAG(m) to send the message m (whose type is TAG) to its neighbors. It is assumed that a process invokes bcast() only at a beginning of a time slot (round). When a message TAG(m) arrives at a process p_i, this process is immediately warned of it, which triggers the execution of the operation receive() to obtain and process the message. Hence, a message is always received and processed during the time slot –round– in which it was broadcast.

From a linguistic point of view, we use the two following **when** notations when writing algorithms, where **predicate** is a predicate involving $CLOCK$ and possibly local variables of the concerned process.

when TAG(m) **is received do** communication-free processing of the message.
when predicate **do** code entailing at most one bcast() invocation.

Message collision and message conflict in the m-bounded memory model. As announced in the Introduction, there is no dedicated communication medium for each pair of communicating processes, and each process has local communication and memory constraints such that, at every round, it cannot receive messages from more than m of it neighbors. If communication is not controlled, "message clash" problems can occur, messages corrupting each other. Consider a process p_i these problems are the following.

- If more than m neighbors of p_i invoke the operation bcast() during the same time slot (round), a message *collision* occurs.
- If p_i and one of its neighbors invoke bcast() during the same time slot (round), a message *conflict* occurs.

As indicated in the introduction, an aim of coloring is to prevent message clashes from occurring, i.e., in our case, ensures C2m-freedom. Let us observe that a coloring algorithm must itself be C2m-free.

3   Coloring with Communication/Memory Constraints

Definition of the CCMC problem. Let $\{p_1, \cdots, p_n\}$ be the n vertices of a connected undirected graph. As already indicated, *neighbors$_i$* denotes the set of the neighbors of p_i. Let the color domain be the set of non-negative integers, and m and K be two positive integers. The aim is to associate a set of colors, denoted *colors$_i$*, with each vertex p_i, such that the following properties are satisfied.

- Conflict-freedom. $\forall i, j : (p_i$ and p_j are neighbors$) \Rightarrow colors_i \cap colors_j = \emptyset$.
- m-Collision-freedom. $\forall i, \forall c : |\{j : p_j \in neighbors_i \wedge c \in colors_j\}| \le m$.
- Efficiency. $|\cup_{1 \le i \le n} colors_i| \le K$.

The first property states the fundamental property of vertex coloring, namely, any two neighbors are assigned distinct colors sets. The second property states the m-constraint coloring on the neighbors of every process, while the third property states an upper bound on the total number of colors that can be used.

As indicated in the Introduction, this problem is denoted CCMC$(n, m, K, 1)$ if each color set is constrained to be a singleton, and CCMC$(n, m, K, \ge 1)$ if there is no such restriction.

Example. An example of such a multi-coloring of a 21-process network, where $\Delta = 10$, and with the constraint $m = 3$, is given in Fig. 1. Notice that $K = \lceil \frac{\Delta}{m} \rceil + 1 = 5$ (the color set is $\{0, 1, 2, 3, 4\}$).

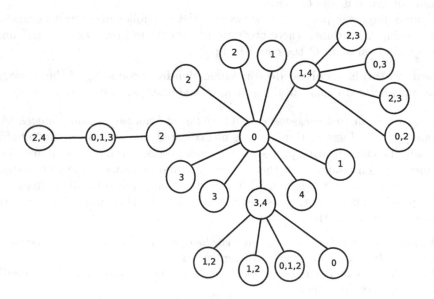

Fig. 1. Multi-coloring of a 21-process 10-degree tree with the constraint $m = 3$

Particular instances. The problem instance $CCMC(n, \infty, K, 1)$ is nothing other than the classical vertex coloring problem, where at most K different colors are allowed ($m = \infty$ states that no process imposes a constraint on the colors of its neighbors, except that they must be different from its own color). The problem instance $CCMC(n, 1, K, 1)$ is nothing other than the classical distance-2 coloring problem (vertices at distance ≤ 2 have different colors).

Using the colors. The reader can easily see that $CCMC(n, m, K, \geq 1)$ captures the general coloring problem informally stated in the introduction. Once a process p_i has been assigned a set of colors $colors_i$, at the application programming level, it is allowed to broadcast a message to neighbors at the rounds (time slots) corresponding to the values of $CLOCK$ such that $(CLOCK \bmod K) \in colors_i$.

4 $CCMC(n, m, K, \geq 1)$ in a Tree Network: Lower Bounds

4.1 An Impossibility Result

Theorem 1. *Neither* $CCMC(n, m, K, 1)$, *nor* $CCMC(n, m, K, \geq 1)$ *can be solved when* $K \leq \lceil \frac{\Delta}{m} \rceil$.

Proof. Let us first show that there is no algorithm solving $CCMC(n, m, K, 1)$ when $K \leq \lceil \frac{\Delta}{m} \rceil$. To this end, let us consider a process p_ℓ, which has Δ neighbors (by the very definition of Δ, there is a such process). Let $\Delta = m \times x + y$, where $0 \leq y < m$. Hence, $x = \frac{\Delta - y}{m} = \lfloor \frac{\Delta}{m} \rfloor$ colors are needed to color $\Delta - y = m \times x$ processes. Moreover, if $y \neq 0$, one more color is needed to color the $y < m$ remaining processes. It follows that $\lceil \frac{\Delta}{m} \rceil$ is a lower bound to color the neighbors of p_ℓ. As p_ℓ cannot have the same color as any of its neighbors, it follows that at least $\lceil \frac{\Delta}{m} \rceil + 1$ are necessary to color $\{p_i\} \cup neighbors_i$, which proves the theorem for $CCMC(n, m, K, \geq 1)$.

Let us observe that an algorithm solving $CCMC(n, m, K, 1)$ can be obtained from an algorithm solving $CCMC(n, m, K, \geq 1)$ by associating with each p_i a single color of its set $colors_i$. Hence, any algorithm solving $CCMC(n, m, \lceil \frac{\Delta}{m} \rceil, \geq 1)$ can be used to solve $CCMC(n, m, \lceil \frac{\Delta}{m} \rceil, 1)$. As $CCMC(n, m, \lceil \frac{\Delta}{m} \rceil, 1)$ is impossible to solve, it follows that $CCMC(n, m, \lceil \frac{\Delta}{m} \rceil, \geq 1)$ is also impossible to solve. □

4.2 A Necessary and Sufficient Condition for Multicoloring

Let $CCMC(n, m, \lceil \frac{\Delta}{m} \rceil + 1, > 1)$ denote the problem $CCMC(n, m, \lceil \frac{\Delta}{m} \rceil + 1, \geq 1)$ where at least one node obtains more than one color.

Theorem 2. $CCMC(n, m, \lceil \frac{\Delta}{m} \rceil + 1, > 1)$ *can be solved on a tree of maximal degree* Δ, *if and only if* (Proof in [18])

$$\exists i : \lceil \frac{\Delta}{m} \rceil + 1 > \max \left(\{ \lceil \frac{\Delta_i}{m} \rceil \} \cup \left\{ \lfloor \frac{\Delta_j}{m} \rfloor \mid p_j \in neighbors_i \right\} \right) + 1.$$

5   CCMC($n, m, K, \geq 1$) in Tree Networks: Algorithm

The algorithm presented in this section use as a skeleton a parallel traversal of a tree [24]. Such a traversal is implemented by control messages that visit all the processes, followed by a control flow that returns at the process that launched the tree traversal.

Algorithm 1 is a C2m-free algorithm that solves the CCMC($n, m, \lceil \frac{\Delta}{m} \rceil +$ $1, \geq$ 1) problem. It assumes that a single process initially receives an external message START(), which dynamically defines it as the root of the tree. This message and the fact that processes at distance smaller or equal to 2 do not have the same identity provide the initial asymmetry from which a deterministic coloring algorithm can be built. The reception of the message START() causes the receiving process (say p_r) to simulate the reception of a fictitious message COLOR(), which initiates the sequential traversal.

Messages. The algorithm uses two types of messages, denoted COLOR() and TERM().

- The messages COLOR() implement a control flow visiting in parallel the processes of the tree from the root to the leaves. Each of them carries three values, denoted *sender*, *cl_map*, and *max_cl*.
  - *sender* is the identity of the sender of the message. If it is the first message COLOR() received by a process p_i, *sender* defines the parent of p_i in the tree.
  - *cl_map* is a dictionary data structure with one entry for each element in *neighbors$_x$* $\cup$ $\{id_x\}$, where p_x is the sender of the message COLOR(). *cl_map*$[id_x]$ is the set of colors currently assigned to the sender and, for each $id_j \in$ *neighbor$_x$*, *cl_map*$[id_j]$ is the set of colors that p_x proposes for p_j.
  - *max_cl* is an integer defining the color domain used by the sender, namely the color set $\{0, 1, \ldots, (max\_cl - 1)\}$. Each child p_i of the message sender will use the color domain defined by $\max(max\_cl, \sigma_i)$ to propose colors to its own children (σ_i is defined below). Moreover, all the children of the sender will use the same slot span $\{0, 1, \ldots, (max\_cl - 1)\}$ to broadcast their messages. This ensures that their message broadcasts will be collision-free[5].
- The messages TERM() propagate the return of the control flow from the leaves to the root. Each message TERM() carries two values: the identity of the destination process (as this message is broadcast, this allows any receiver to know if the message is for it), and the identity of the sender.

---

[5] As we will see, conflicts are prevented by the message exchange pattern imposed by the algorithm.

Local variables. Each process p_i manages the following local variables. The constant $\Delta_i = |neighbors_i|$ is the degree of p_i, while the constant $\sigma_i = \lceil \frac{\Delta_i}{m} \rceil + 1$ is the number of colors needed to color the star graph made up of p_i and its neighbors.

- *state$_i$* (initialized to 0) is used by p_i to manage the progress of the tree traversal. Each process traverses five different states during the execution of the algorithm. States 1 and 3 are active states: a process in state 1 broadcasts a COLOR() message for its neighbors, while a process in state 3 broadcasts a message TERM() which has a meaning only for its parent. States 0 and 2 are waiting states in which a process listens on the broadcast channels but cannot send any message. Finally, state 4 identifies local termination.
- *parent$_i$* stores the identity of the process p_j from which p_i receives a message COLOR() for the first time (hence p_j is the parent of p_i in the tree). The root p_r of the tree, defined by the reception of the external message START(), is the only process such that $parent_r = id_r$.
- *colored$_i$* is a set containing the identities of the neighbors of p_i that have been colored.
- *to_color$_i$* is the set of neighbors to which p_i must propagate the coloring (network traversal).
- *color_map$_i$[neighbors$_i$ $\cup$ {id$_i$}]* is a dictionary data structure where p_i stores the colors of its neighbors in *color_map$_i$[neighbors$_i$]*, and its own colors in *color_map$_i$[id$_i$]*; *colors$_i$* is used as a synonym of *color_map$_i$[id$_i$]*.
- *max_cl$_i$* defines both the color domain from which p_i can color its children, and the time slots (rounds) at which its children will be allowed to broadcast.
- *slot_span$_i$* is set to the value *max_cl* carried by the message COLOR() received by p_i from its parent. As this value is the same for all the children of its parent, they will use the same slot span to define the slots during which each child will be allowed to broadcast messages.

Initial state. In its initial state ($state_i = 0$), a process p_i waits for a message COLOR(). As already indicated, a single process receives the external message START(), which defines it at the root process. It is assumed that $CLOCK = 0$ when a process receives this message. When it receives it, the corresponding process p_i simulates the reception of the message COLOR(id_i, cl_map, σ_i) where $cl\_map[id_i]$ defines its color, namely, $(CLOCK + 1) \bmod \sigma_i$ (lines 2–3). Hence, at round number 1, the root will send a message COLOR() to its children (lines 26–27).

Algorithm: reception of a message COLOR(). When a process p_i receives a message COLOR() for the first time, it is visited by the network traversal, and must consequently (a) obtain an initial color set, and (b) propagate the network traversal, if it has children. The processing by p_i of this first message COLOR($sender$, cl_map, max_cl) is done at lines 6–24. First, p_i saves the identity of its parent (the sender of the message) and its proposed color set (line 6), initializes *colored$_i$* to {$sender$}, and *to_color$_i$* to its other neighbors (line 7). Then

---

1 **Init:** $\sigma_i = \lceil \frac{\Delta_i}{m} \rceil + 1$; $state_i \leftarrow 0$; $colors_i \leftarrow \emptyset$; $colors_i$ stands for $color\_map_i[id_i]$

2 **when** START() **is received do**   ▷ *A single process p_i receives this message.*

3 | p_i executes lines 5-25 as if it received the message COLOR(id_i, cl_map, σ_i) where $cl\_map[id_i] = \{(CLOCK + 1) \bmod \sigma_i\}$

4 **when** COLOR($sender, cl\_map, max\_cl$) **is received do**

5 | **if** *first message* COLOR() *received* **then**

6 | | $parent_i \leftarrow sender$; $color\_map_i[parent_i] \leftarrow cl\_map[sender]$

7 | | $colored_i \leftarrow \{sender\}$; $to\_color_i \leftarrow neighbors_i \setminus \{sender\}$

8 | | $color\_map_i[id_i] \leftarrow cl\_map[id_i]$ ▷ *Synonym of $colors_i$*

9 | | $max\_cl_i \leftarrow \mathsf{max}(max\_cl, \sigma_i)$; $slot\_span_i \leftarrow max\_cl$

10 | | **if** $(to\_color_i \neq \emptyset)$ **then**   ▷ *next lines: $tokens_i$ is a multiset.*

11 | | | $tokens_i \leftarrow \{m \text{ tokens with color } x, \forall x \in ([0..(max\_cl_i - 1)] \setminus colors_i)\}$
                     $\setminus \{1 \text{ token with color } z, \forall z \in color\_map_i[parent_i]\}$

12 | | | **while** $(|tokens_i| < |to\_colors_i|)$ **do**

13 | | | | **if** $(|colors_i| > 1)$ **then**

14 | | | | | **let** $cl \in colors_i$; suppress cl from $colors_i$

15 | | | | | add m tokens colored cl to $tokens_i$

16 | | | | **else**

17 | | | | | **let** cl be the maximal color in $color\_map_i[parent_i]$

18 | | | | | add one token colored cl to $tokens_i$

19 | | | | | $color\_map_i[parent_i] \leftarrow color\_map_i[parent_i] \setminus \{cl\}$

20 | | | Extract $|to\_colors_i|$ non-empty non-intersecting multisets $tk[id]$ (where $id \in to\_color_i$) from $tokens_i$ such that no $tk[id]$ contains several tokens with the same color

21 | | | **foreach** $id \in to\_color_i$ **do**

22 | | | | $color\_map_i[id] \leftarrow \{\text{colors of the tokens in } tk[id]\}$

23 | | | $state_i \leftarrow 1$ ▷ *p_i has children*

24 | | **else** $state_i \leftarrow 3$ ▷ *p_i is a leaf*

25 | **else** $color\_map_i[id_i] \leftarrow color\_map_i[id_i] \cap cl\_map[id_i]$

26 **when** $((CLOCK \bmod slot\_span_i) \in colors_i) \wedge (state_i \in \{1,3\}))$ **do**

27 | **case** $(state_i = 1)$ bcast COLOR($id_i, color\_map_i, max\_cl_i$) ; $state_i \leftarrow 2$

28 | **case** $(state_i = 3)$ bcast TERM($parent_i, id_i$); $state_i \leftarrow 4$ ▷*p_i's subtree is done*

29 **when** TERM($dest, id$) **is received do**

30 | **if** $(dest \neq id_i)$ **then** discard the message (do not execute lines 31-34)

31 | $colored_i \leftarrow colored_i \cup \{id\}$

32 | **if** $colored_i = neighbors_i$ **then**

33 | | **if** $parent_i = id_i$ **then** the root p_i claims termination

34 | | **else** $state_i \leftarrow 3$

Algorithm 1. C2m-free algorithm solving CCMC($n, m, \lceil \frac{\Delta}{m} \rceil + 1, \geq 1$) in tree networks (code for p_i)

p_i obtains a color set proposal from the dictionary cl_map carried by the message (line 8), computes the value max_cl_i from which its color palette will be defined, and saves the value max_cl carried by the message COLOR() in the local variable $slot\_span_i$ (line 9). Let us remind that the value max_cl_i allows it to know the color domain used up to now, and the rounds at which it will be able to broadcast messages (during the execution of the algorithm) in a collision-free way.

Then, the behavior of p_i depends on the value of to_color_i. If to_color_i is empty, p_i is a leaf, and there is no more process to color from it. Hence, p_i proceeds to state 3 (line 24).

If to_color_i is not empty, p_i has children. It has consequently to propose a set of colors for each of them, and save these proposals in its local dictionary $color\_map_i[neighbors_i]$. To this end, p_i computes first the domain of colors it can use, namely, the set $\{0, 1, \ldots, (max\_cl_i - 1)\}$, and considers that each of these colors c is represented by m tokens colored c. Then, it computes the multiset[6], denoted $tokens_i$, containing all the colored tokens it can use to build a color set proposal for each of its children (line 11). The multiset $tokens_i$ is initially made up of all possible colored tokens, from which are suppressed (a) all tokens associated with the colors of p_i itself, and, (b) one colored token for each color in $color\_map_i[parent_i]$ (this is because, from a coloring point of view, its parent was allocated one such colored token for each of its colors).

Then, p_i checks if it has enough colored tokens to allocate at least one colored token to each of its children (assigning thereby the color of the token to the corresponding child). If the predicate $|tokens_i| \geq |to\_color_i|$ is satisfied, p_i has enough colored tokens and can proceed to assign sets of colors to its children (lines 20–22). Differently, if the predicate $|tokens_i| < |to\_color_i|$ is satisfied, p_i has more children than colored tokens. Hence, it must find more colored tokens. For that, if $colors_i$ (i.e., $color\_map_i[id_i]$) has more than one color, p_i suppresses one color from $colors_i$, adds the m associated colored tokens to the multiset $tokens_i$ (lines 14–15), and re-enters the "while" loop (line 12). If $colors_i$ has a single color, this color cannot be suppressed from $colors_i$. In this case, p_i considers the color set of its parent ($color\_map_i[parent_i]$), takes the maximal color of this set, suppresses it from $color\_map_i[parent_i]$, adds the associated colored token to the multiset $tokens_i$, and –as before– re-enters the "while" loop (line 16–19). Only one token colored cl is available because the $(m - 1)$ other tokens colored cl were already added into the multiset $tokens_i$ during its initialization at line 11.

As already said, when the predicate $|tokens_i| < |to\_color_i|$ (line 12) becomes false, $tokens_i$ contains enough colored tokens to assign to p_i's children. This assignment is done at lines 20–22. Let $ch = |to\_color_i|$ (number of children of p_i); p_i extracts ch pairwise disjoint and non-empty subsets of the multiset $tokens_i$, and assigns each of them to a different neighbor. "Non-empty non-intersecting

---

[6] Differently from a set, a *multiset* (also called a *a bag*), can contain several times the same element. Hence, while $\{a, b, c\}$ and $\{a, b, a, c, c, c\}$ are the same set, they are different multisets.

multisets" used at line 20 means that, if each of z multisets $tk[id_x]$ contains a token with the same color, this token appears at least z times in $tokens_i$.

If the message COLOR($sender, cl\_map, -$) received by p_i is not the first one, it was sent by one of its children. In this case, p_i keeps in its color set $color\_map_i[id_i]$ ($colors_i$) only colors allowed by its child $sender$ (line 25). Hence, when p_i has received a message COLOR() from each of its children, its color set $colors_i$ has its final value.

Algorithm: broadcast of a message. A process p_i is allowed to broadcast a message only at the rounds corresponding to a color it obtained (a color in $colors_i = color\_map_i[id_i]$ computed at lines 8, 14, and 25), provided that its current local state is 1 or 3 (line 26).

If $state_i = 1$, p_i received previously a message COLOR(), which entailed its initial coloring and a proposal to color its children (lines 11–23). In this case, p_i propagates the tree traversal by broadcasting a message COLOR() (line 27), which will provide each of its children with a coloring proposal. Process p_i then progresses to the local waiting state 2.

If $state_i = 3$, the coloring of the tree rooted at p_i is terminated. Process p_i consequently broadcasts TERM($parent_i, id_i$) to inform its parent of it. It also progresses from state 3 to state 4, which indicates its local termination (line 28).

Algorithm: reception of a message TERM(). When a process p_i receives such a message it discards it if it is not the intended destination process (line 30). If the message is for it, p_i adds the sender identity to the set $colored_i$ (line 31). Finally, if $colored_i = neighbors_i$, p_i learns that the subtree rooted at it is colored (line 32). It follows that, if p_i is the root ($parent_i = i$), it learns that the algorithm terminated. Otherwise, it enters state 3, that will direct it to report to its parent the termination of the coloring of the subtree rooted at it.

Solving CCMC(n,m,K,1) in a tree. Algorithm 1 can be easily modified to solve CCMC($n, m, K, 1$). When a process enters state 3 (at line 24 or line 34), it reduces $color\_map_i[id_i]$ (i.e., $colors_i$) to obtain a singleton.

6 CCMC($n, m, K, \geq 1$) in a Tree: Cost and Proof

The proof assumes $n > 1$. Let us remember that $colors_i$ and $color\_map_i[id_i]$ are the same local variable of p_i, and p_r denotes the dynamically defined root.

Cost of the algorithm. Each non-leaf process broadcasts one message COLOR(), and each non-root process broadcasts one message TERM(). Let x be the number of leaves. There are consequently $(2n - (x+1))$ broadcasts. As $\Delta \leq x+1$, the number of broadcast is upper bounded by $2n - \Delta$.

Given an execution whose dynamically defined root is the process p_r, let d be the height of the corresponding tree. The root computes the colors defining the slots (rounds) at which its children can broadcast the messages COLOR()

and TERM(). These colors span the interval $[0..\lceil\frac{\Delta_r}{m}\rceil]$, which means that the broadcasts of messages COLOR() by the processes at the first level of the tree span at most $\lceil\frac{\Delta_r}{m}\rceil + 1$ rounds. The same broadcast pattern occurs at each level of the tree. It follows that the visit of the tree by the messages COLOR() requires at most $O(d\lceil\frac{\Delta}{m}\rceil)$ rounds. As the same occurs for the messages TERM(), returning from the leaves to the root, it follows that the time complexity is $O(d\lceil\frac{\Delta}{m}\rceil)$.

Proof of the algorithm

Theorem 3. *Let* $K = \lceil\frac{\Delta}{m}\rceil + 1$. *Algorithm 1 is a C2m-free algorithm, which solves* CCMC$(n, m, K, \geq 1)$ *in tree networks. Moreover, it is optimal with respect to the value of* K.

The proof is decomposed into nine lemmas showing that the algorithm (a) is itself conflict-free and collision-free, (b) terminates, and (c) associates with each process p_i a non-empty set $colors_i$ satisfying the Conflict-freedom, m-Collision-freedom and Efficiency properties defined in Sect. 3.

To this end, a notion of *well-formedness* suited to COLOR() messages is introduced. More precisely a message COLOR($sender, cl\_map, max\_cl$) is well-formed if its content satisfies the following properties. Let $sender = id_i$.

M1 The keys of the dictionary cl_map are the identities in $neighbors_i \cup \{id_i\}$.
M2 $\forall\, id \in (neighbors_i \cup \{id_i\}) :\ cl\_map[id] \neq \emptyset$.
M3 $\forall\, id \in neighbors_i :\ cl\_map[id] \cap cl\_map[id_i] = \emptyset$.
M4 $\forall c :\ |\{j : (id_j \in neighbors_i) \wedge (c \in cl\_map[id_j])\}| \leq m$.
M5 $1 < max\_cl \leq \lceil\frac{\Delta}{m}\rceil + 1$.
M6 $\forall\, id \in (neighbors_i \cup \{id_i\}) :\ cl\_map[id] \subseteq [0..max\_cl - 1]$.

Due to page limitation, the missing lemmas are proved in [18]. Here we only give the proof of one them.

Lemma 1. *If a process* p_i *computes a color set* ($colors_i$), *this set is not empty.*

Proof. Let us first observe that, if a process $p_i \neq p_r$ receives a message COLOR($-, cl\_map, -$), the previous lemma means that this message is well-formed, and due to property M2, its field $cl\_map[id_i]$ is not empty, from which follows that the initial assignment of a value to $color\_map_i[id_i] \equiv colors_i$ is a non-empty set. Let us also observe, that, even if it is not well-formed the message COLOR($-, cl\_map, -$) received by the root satisfies this property. Hence, any process that receives a message COLOR() assigns first a non-empty value to $color\_map_i[id_i] \equiv colors_i$.

Subsequently, a color can only be suppressed from $color\_map_i[id_i] \equiv colors_i$ at line 25 when p_i receives a message COLOR() from one of its children. If p_i is a leaf, it has no children, and consequently never executes line 25. So, let us assume that p_i is not a leaf and receives a message COLOR($id_j, cl\_map, -$) from one of its children p_j. In this case p_i previously broadcast at line 27 a message COLOR($id_i, color\_map_i, -$) that was received by p_j and this message is well-formed (this is proved in another lemma).

A color c that is suppressed at line 25 when p_i processes COLOR(id_j, cl_map, $-$) is such that $c \in colors_i$ and $c \notin cl\_map[id_i]$. $cl\_map[id_i]$ can be traced back to the local variable $color\_map_j[id_i]$ used by p_j to broadcast COLOR() at line 27. Tracing the control flow further back, $color\_map_j[id_i]$ was initialized by p_j to $color\_map_i[id_i]$ (line 6) when p_j received the well-formed message COLOR() from p_i. When processing COLOR() received from p_i, process p_j can suppress colors from $color\_map_j[id_i]$ only at line 19, where it suppresses colors starting from the greatest remaining color. We have the following.

- If p_i is not the root, the message COLOR() it received was well-formed (this is proved in another lemma). In this case, it follows from a lemma proved in [18] that it always remains at least one color in $color\_map_j[id_i]$.
- If $p_i = p_r$, its set $colors_r$ is a singleton (it "received" COLOR(id_r, cl_map_r, $-$) where cl_map_r has a single entry, namely $cl\_map_r[id_r] = \{1\}$).
  When p_j computes $tokens_j$ (line 11) we have
  $$|tokens_j| = m \times \max(\sigma_r, \sigma_j) - m = m\lceil \tfrac{\max(\Delta_r, \Delta_j)}{m} \rceil \geq \max(\Delta_r, \Delta_j) \geq \Delta_j = |to\_colors_j|,$$
  from which follows that $|tokens_j| \geq |to\_colors_j| = |neighbors_j| - 1$. Hence, p_j does not execute the loop, and consequently does not modify $color\_map_j[id_r]$.

Consequently, the smallest color of $colors_i \equiv color\_map_i[id_i]$ is never withdrawn from $color\_map_j[id_i]$. It follows that, at line 25, p_i never withdraws its smallest color from the set $color\_map_i[id_i]$.   □

7 Conclusion

The paper first introduced a new vertex coloring problem (called CCMC), in which a process may be assigned several colors in such a way that no two neighbors share colors, and for any color c, at most m neighbors of any vertex share the color c. This coloring problem is particularly suited to assign rounds (slots) to processes (nodes) in broadcast/receive synchronous communication systems with communication or local memory constraints. Then, the paper presented a distributed algorithm which solve this vertex coloring problem for tree networks in a round-based programming model with conflicts and (multi-frequency) collisions. This algorithm is optimal with respect to the total number of colors that can be used, namely it uses only $K = \lceil \frac{\Delta}{m} \rceil + 1$ different colors, where Δ is the maximal degree of the graph.

It is possible to easily modify the coloring problem CCMC to express constraints capturing specific broadcast/receive communication systems. As an example, suppressing the conflict-freedom constraint and weakening the collision-freedom constraint

$$\forall i, \forall c : |\{j : (id_j \in neighbors_i \cup \{id_i\}) \wedge (c \in colors_j)\}| \leq m, \qquad (1)$$

captures bi-directional communication structures encountered in some practical systems in which nodes may send and receive on distinct channels during the same round. Interestingly, solving the coloring problem captured by (1) is equivalent to solving distance-2 coloring in the sense that a purely local procedure (i.e., a procedure involving no communication between nodes) executed on each node can transform a classical distance-2 coloring into a multi-coloring satisfying (1). More precisely, assuming a coloring $\text{col} : V \mapsto [0..(K * m) - 1]$ providing a distance-2 coloring with $K * m$ colors on a graph $G = (V, E)$, it is easy to show that the coloring (with one color per vertex)

$$\begin{aligned} \text{col}' : V &\mapsto [0 .. K - 1] \\ x &\to \text{col}(x) \text{ mod } K, \end{aligned} \tag{2}$$

fulfills (1) on G [7]. Since the distance-2 problem with $K * m$ colors is captured by $\text{CCMC}(n, 1, K * m, 1)$ (as discussed in Sect. 3), the proposed algorithm can also solve the coloring condition captured by (1) on trees in our computing model.

Moreover, from an algorithmic point of view, the proposed algorithm is versatile, making it an attractive starting point to address other related problems. For instance, in a heterogeneous network, lines 20–22 could be modified to take into account additional constraints arising from the capacities of individual nodes, such as their ability to use only certain frequencies.

Last but not least, a major challenge for future work consists in solving the CCMC problem in general graphs. The new difficulty is then to take into account cycles.

Acknowledgments. This work has been partially supported by the Franco-German DFG-ANR Project 40300781 DISCMAT (devoted to connections between mathematics and distributed computing), and the Franco-Hong Kong ANR-RGC Joint Research Programme 12-IS02-004-02 CO2Dim.

References

1. Angluin, D.: Local and global properties in networks of processors. In: Proceedings of 12th ACM Symposium on Theory of Computation (STOC 1981), pp. 82–93. ACM Press (1981)
2. Attiya, H., Welch, J.: Distributed Computing: Fundamentals, Simulations and Advanced Topics, 2nd edn. Wiley-Interscience, Hoboken (2004)
3. Barenboim, L., Elkin, M.: Deterministic distributed vertex coloring in polylogarithmic time. J. ACM **58**(5), 23 (2011)
4. Barenboim, L., Elkin, M.: Distributed Gaph Coloring, Fundamental and Recent Developments, p. 155. Morgan & Claypool Publishers (2014)
5. Barenboim, L., Elkin, M., Kuhn, F.: Distributed (Delta+1)-coloring in linear (in Delta) time. SIAM J. Comput. **43**(1), 72–95 (2014)
6. Blair, J., Manne, F.: An efficient self-stabilizing distance-2 coloring algorithm. In: Kutten, S., Žerovnik, J. (eds.) SIROCCO 2009. LNCS, vol. 5869, pp. 237–251. Springer, Heidelberg (2010). doi:10.1007/978-3-642-11476-2_19

---

[7] This is because (a) distance-2 coloring ensures that any vertex and its neighbors have different colors, and (b) there are at most m colors $c_1, ..., c_x \in [0..(K * m) - 1]$ (hence $x \leq m$), such that $(c_1 \text{ mod } K) = \cdots = (c_x \text{ mod } K) = c \in [0..(K - 1)]$.

7. Bozdağ, D., Catalyurek, U., Gebremedhin, A.H., Manne, F., Boman, E.G., Özgüner, F.: A parallel distance-2 graph coloring algorithm for distributed memory computers. In: Yang, L.T., Rana, O.F., Martino, B., Dongarra, J. (eds.) HPCC 2005. LNCS, vol. 3726, pp. 796–806. Springer, Heidelberg (2005). doi:10.1007/11557654_90
8. Bozdag, D., Gebremedhin, A.S., Manne, F., Boman, G., Çatalyürek, U.V.: A framework for scalable greedy coloring on distributed-memory parallel computers. J. Parallel Distrib. Comput. **68**(4), 515–535 (2008)
9. Chipara, O., Lu, C., Stankovic, J., Roman, G.-C.: Dynamic conflict-free transmission scheduling for sensor network queries. IEEE Trans. Mobile Comput. **10**(5), 734–748 (2011)
10. Cole, R., Vishkin, U.: Deterministic coin tossing with applications to optimal parallel list ranking. Inf. Control **70**(1), 32–53 (1986)
11. Frey, D., Lakhlef, H., Raynal, M.: Optimal collision/conflict-free distance-2 coloring in synchronous broadcast/receive tree networks. Research Report (2015). https://hal.inria.fr/hal-01248428
12. Garey, M.R., Johnson, D.S.: Computers and Intractability: A Guide to the Theory of NP-Completeness. W.H. Freeman & Co., New York (1979)
13. Gebremedhin, A.H., Manne, F., Pothen, A.: Parallel distance-k coloring algorithms for numerical optimization. In: Monien, B., Feldmann, R. (eds.) Euro-Par 2002. LNCS, vol. 2400, pp. 912–921. Springer, Heidelberg (2002). doi:10.1007/3-540-45706-2_130
14. Goldberg, A., Plotkin, S., Shannon, G.: Parallel symmetry-breaking in sparse graphs. SIAM J. Disc. Math. **1**(4), 434–446 (1988)
15. Herman, T., Tixeuil, S.: A sistributed TDMA slot assignment algorithm for wireless sensor networks. In: Nikoletseas, S.E., Rolim, J.D.P. (eds.) ALGOSENSORS 2004. LNCS, vol. 3121, pp. 45–58. Springer, Heidelberg (2004). doi:10.1007/978-3-540-27820-7_6
16. Hugh, H.H., Molloy, M., Reed, B.: Colouring a graph frugally. Combinatorica **17**(4), 469–482 (1997)
17. Kuhn, F., Wattenhofer, R.: On the complexity of distributed graph coloring. In: Proceedings of 25th ACM Symposium Principles of Distributed Computing (PODC 2006), pp. 7–15. ACM Press (2006)
18. Lakhlef, H., Raynal, M., Taïani, F.: Vertex coloring with communication and local memory constraints in synchronous broadcast networks. Tech report 2035, p. 24. IRISA, Université de Rennes (F) (2016). https://hal.inria.fr/hal-01300095
19. Linial, N.: Locality in distributed graph algorithms. SIAM J. Comput. **21**(1), 193–201 (1992)
20. Lynch, N.A.: Distributed Algorithms. Morgan Kaufmann, San Francisco (1996)
21. Molloy, M., Bruce, R.B.: Asymptotically optimal frugal colouring. J. Comb. Theory Ser. B **100**(2), 247–249 (2010). Corrigendum in J. Comb. Theory Ser. B 100(2), 226–246 (2010)
22. Peleg, D.: Distributed Computing, A Locally Sensitive Approach. SIAM Monographs on Discrete Mathematics and Applications, p. 343 (2000). ISBN 0-89871-464-8
23. Raynal, M.: Fault-Tolerant Agreement in Synchronous Message-Passing Systems, p. 165. Morgan & Claypool Publishers (2010). ISBN 978-1-60845-525-6
24. Raynal, M.: Distributed Algorithms for Message-Passing Systems. Springer, Heidelberg (2013). ISBN 978-3-642-38122-5

A New Kind of Selectors
and Their Applications to Conflict Resolution
in Wireless Multichannels Networks

Annalisa De Bonis$^{(\boxtimes)}$ and Ugo Vaccaro

Dipartimento di Informatica, Università di Salerno, 84084 Fisciano, SA, Italy
debonis@dia.unisa.it, uvaccaro@unisa.it

Abstract. We investigate the benefits of using multiple channels of communications in wireless networks, under the full-duplex multi-packet reception model of communication. The main question we address is the following: Is a speedup linear in the number of channels achievable, for some interesting communication primitive? We provide a positive answer to this interrogative for the *Information Exchange Problem*, in which k arbitrary nodes have information they intend to share with the entire network. To achieve this goal, we devise and exploit a combinatorial structure that generalizes well known combinatorial tools widely used in the area of data-exchange in multiple access channels (i.e., strongly selective families, selectors, and related mathematical objects). For our new combinatorial structures we provide both existential results, based on the Lovász Local Lemma, and efficient constructions, leveraging on properties of error correcting codes. We also prove non existential results, showing that our constructions are not too far from being optimal.

1 Introduction

Recent advances in technology and protocols have made wireless devices and services that operate on multiple channels increasingly available (e.g., WiFi [1] and Bluetooth [6]). Furthermore, the introduction of *realistic* full-duplex relaying technology (see [4,29,33] and references therein quoted), that allow terminals to transmit and receive *simultaneously* over a same frequency band, opens up a new spectrum of potentialities to greatly increase the throughput of a communication system. However, technology alone is not necessarily a panacea. More often than not, technological advances need to be supported by the introduction of new conceptual tools in order to transit from a mere potentiality into a full actuality. Under these premises, in this paper we seek an answer to the following essential question: How much faster can we disseminate data in a network if we have such multiple communication channels at our disposal (as opposed to relying on single communication channels)?

1.1 The Communication Model

Our scenario consists of a single-hop communication network in which n stations have a common access to a multichannel $\mathcal{C}$, comprising of $q - 1 \geq 1$ individual

© Springer International Publishing AG 2017
M. Chrobak et al. (Eds.): ALGOSENSORS 2016, LNCS 10050, pp. 45–61, 2017.
DOI: 10.1007/978-3-319-53058-1_4

channels labeled with the integers $1, 2, \ldots, q - 1$ (think of a wireless network, with stations able to transmit and listen over $q - 1$ different frequencies). We assume that at most a certain number $k \leq n$ of stations might be *active* at the same time, i.e., might want to transmit simultaneously over the channels. Each station may choose which of the $q - 1$ channels to use for transmitting. An active station *successfully* transmits if and only if it transmits *singly* on the channel of its choice. To further specify the communication model, we follow the footsteps of the seminal paper by Massey and Mathys [30]. We assume that time is divided into time slots and that transmissions occur during these time slots. We also assume that all stations have a global clock and that active stations start transmitting at the same time slot. A scheduling algorithm for such a multiaccess system is a protocol that schedules the transmissions of the n stations over a certain number t of time slots (*steps*) identified by integers $1, 2, \ldots, t$. In this paper, a *conflict resolution* algorithm for the above described multiaccess system is a scheduling algorithm that allows all active stations to transmit successfully. A *non adaptive* conflict resolution algorithm is a protocol that schedules all transmissions in advance, i.e., for each step $i = 1, \ldots, t$, it establishes which stations should transmit at step i without looking at what happened over the channels at the previous steps. A non adaptive scheduling algorithm can be represented by a set of n vectors, identified by integers from 1 through n (each of which corresponding to a distinct station), of length t over the alphabet $\{0, 1, \ldots, q - 1\}$, with the meaning that station j is scheduled to transmit at step i over the channel $s \in \{1, \ldots, q-1\}$ if and only if the i-th entry of its associated vector j is equal to s. If the i-th entry is equal to 0, the station has to stay silent at step i. A non adaptive scheduling algorithm is conveniently represented by the q-ary matrix having as columns the n q-ary vectors associated with the scheduling of the transmissions of the n stations. Obviously, entry (i, j) of such a matrix is equal to $s \neq 0$ if and only if station j is scheduled to transmit at step i over channel $s \in \{1, \ldots, q - 1\}$, and it is equal to $s = 0$ if and only if station j remains silent at step i. The parameter of interest to be minimized is the number of rows of the matrix. In fact, that number represents the number of time slots in which the conflict resolution algorithm schedules the transmissions of the n stations, so that up to k active stations transmit with success.

In order to fully define the communication model, we must also specify *whether or not* each station is able to discriminate between the event that its transmission has been successful, from the event that its transmission has been unsuccessful (i.e., another station has attempted a simultaneous transmission over the same channel at same time instant, thus causing a collision). In other words, we must differentiate the situation in which stations receive feedback from the channels about whether or not their transmissions have been successful, from the situation in which channels do not provide this feedback.

The Multiple-Access Channel Without Feedback. When stations receive no *feedback* from the channel, then the conflict resolution algorithm must necessarily schedule transmissions in such a way that each active station transmits singly over at least one of the $q - 1$ channels at some step. A conflict resolution algorithm for this model can be represented by a q-ary matrix M with the property that for any k columns of M and for any column $\mathbf{c}$ chosen among these k columns, there exists a row in correspondence of which $\mathbf{c}$ has an entry equal to some $s \in \{1, \ldots, q - 1\}$ and the remaining $k - 1$ columns have entries different from s. When $q = 2$, matrices that satisfy this property have been very well studied in the literature where they are known under different names, such as superimposed codes [25], $(k - 1)$-cover free families [20], k-disjunct codes [16], and strongly selective families [7,10]. To the best of our knowledge, there are no results in the literature for the multichannel scenario considered in this paper.

The Multiple-Access Channel with Feedback. In addition to the situation when stations receive no feedback from the channels, we consider also the communication model in which any transmitting station receives feedback on whether its transmission has been successful or not (see [27] for the case when only one channel is available). In such a model, an active station has the capability to become *inactive* (i.e., to refrain from transmitting) after it has transmitted successfully. As in the previous model, a non adaptive conflict resolution algorithm should guarantee that for each active station there is a step at which it transmits singly over some of the $q - 1$ available channels. However, in this scenario an active station transmits singly to a channel also at time slots where it is scheduled to transmit simultaneously on the same channel with some of the other k stations that were initially active, provided that these other stations transmitted successfully at one of the *previous* steps (and therefore, have in the meantime become inactive). The characterization of q-ary matrices that represent non adaptive conflict resolution algorithms for this different model of communication is less simple than that for the case without feedback, and will be presented in the technical part of this paper. Again, we are not aware of papers that have already studied this problem in the multichannels scenario considered in this paper.

2   Our Results

In this paper we study non adaptive conflict resolution protocols for both the two above described multiple-access models, that is, for multiple access channels with and without feedback. Our main goal is to provide an answer to the following question: Is a speedup linear in the number of channels achievable, for some interesting communication primitive? We provide a positive answer to this interrogative for the basic *Information Exchange Problem* [24,28], in which k arbitrary nodes have information they intend to share with the entire network. In order to study this question, we introduce two new combinatorial structures that consist in a generalization of selectors [14] (and therefore of superimposed codes [25], $(k-1)$-cover free families [20], k-disjunct codes [16], strongly selective families [7,10]) and a generalization of $\mathrm{KG}(k, n)$-codes [27].

Our paper is organized as follows. In Sect. 4 we first introduce a new version of selectors that, in the present no-feedback scenario, correspond to protocols that schedule transmissions so that, for all possible subsets of k active stations, one has that at least m out of k stations are scheduled to transmit singly over at least one channel $s \in \{1, \ldots, q-1\}$. Then, we introduce a new version of $KG(k, n)$-codes that furnish scheduling protocols for the multiple-access multichannels with feedback that allow to solve all conflicts for all possible subsets of up to k active stations. Since the number n of stations in the network and the number k of active stations can be arbitrary, while the number $q-1$ of channels is usually severely limited by technological factors [1,6], in this paper we assume $q \leq k$. In the full version of this paper we will present our combinatorial results in their full mathematical generality. Due to space limit, some of the proofs are omitted in this version of the paper.

Our main results are summarized by the following two theorems.

Theorem 1. *Let k, m, q, and n be integers such that $1 \leq m \leq k \leq n$, and $2 \leq q \leq k$. There exists a conflict resolution algorithm for a multiple-access channel $\mathcal{C}$ without feedback (comprising of $q-1 \geq 1$ individual channels) that schedules the transmissions of n stations in such a way that, for all possible subsets of up to k active stations, one has that at most $k - m$ out of k active stations fail to transmit successfully. The number t of time slots used by the conflict resolution algorithm is*

$$t \leq \frac{4(k-1)}{(k-m+1)(q-1)} \left(\ln \binom{n}{k-1} + \ln k + \ln \binom{k}{k-m+1} + 1 \right)$$
$$= O\left(\frac{k^2}{(k-m+1)q} \log \frac{n}{k} \right). \tag{1}$$

When $m = k$ (as in the Information Exchange Problem) the number of time slots t used by the conflict resolution algorithm is

$$t = O\left(\frac{k^2}{q} \log \frac{n}{k} \right), \tag{2}$$

and we can prove that for any *conflict resolution algorithm for the Information Exchange Problem it holds that the number of time slots t is such that*

$$t = \Omega\left(\frac{k^2}{q \log k} \log \frac{n}{k} \right). \tag{3}$$

The proof technique of (1) is based on the Lovász Local Lemma and the recent constructive version of it given by Moser and Tardos [32]. To prove (3) we uncover a relation between the combinatorial structures we introduce in this paper and the well known superimposed codes of [25]. In case of channels with feedback, we prove the following (optimal) result.

Theorem 2. *Let k, q, and n be integers such that $2 \leq q \leq k \leq n$. There exists a conflict resolution algorithm for a multiple-access channel $\mathcal{C}$ with feedback* *(comprising of $q - 1 \geq 1$ individual channels) that schedules the transmissions of n stations in such a way that, for all possible subsets of k stations, one has that all active stations transmit successfully. The number t of time slots used by the conflict resolution algorithm is*

$$t = O\left(\frac{k}{q} \log \frac{n}{k}\right). \tag{4}$$

Moreover, for any conflict resolution algorithm in this scenario it holds that the number of time slots t is such that

$$t = \Omega\left(\frac{k}{q} \log \frac{n}{k}\right). \tag{5}$$

We remark that the asymptotic upper bound of Theorem 2 holds also in the case when there is no a priori knowledge of the number of active stations. Indeed, in this case, conflicts are resolved by running iteratively the conflict resolution algorithm given for the case when an upper bound on k is known in advance. At each iteration (stage), the algorithm doubles the number of stations that are *assumed* to be active. In other words, at stage i the conflict resolution algorithm for the bounded case is run for a number k_i of supposedly active stations equal to 2^i. Since at stage $\lceil \log k \rceil$, that algorithm is run for a number of active stations larger than or equal to k, one is guaranteed that all active stations transmit with success within that stage.

In the final sections of this paper we highlight a few interesting connections between the combinatorial structures introduced in this paper and the well known Frameproof Codes [5]. Moreover, we also show the remarkable fact that, for an infinite set of the relevant parameters n and k, one can construct our combinatorial structures in polynomial time and of optimal (minimum) length.

3 Related Work

Information dissemination in single-channel, single-hop networks is a well known and widely studied problem. We refer the reader to the survey papers [9,22] for a presentation of this important area. The study of distributed algorithms in multi-channels wireless networks is relatively recent (see [8,13,15,23,24,39–41] and references therein quoted). However, most of these papers considered either randomized algorithms, or different communication primitives, or different communication models. To the best of our knowledge, our paper seems to be the first to consider the Information Exchange Problem in the communication model described in [4,33].

4 Combinatorial Tools

Let M be an arbitrary t-rows n-columns matrix, where we denote by $\mathbf{c}_1, \ldots, \mathbf{c}_n$ the $t \times 1$ columns of M. We assume that the entries of M are over the alphabet

$\{0, 1, \ldots, q-1\}$, where q is an integer larger than or equal to 2. For each column $\mathbf{c}$ of M and row index $i \in \{1, \ldots, t\}$, we denote by $\mathbf{c}(i)$ the i-th component of vector $\mathbf{c}$. Given arbitrary $\mathbf{c}, \mathbf{c}_1, \ldots, \mathbf{c}_g$ column vectors, we say that $\mathbf{c}$ is *covered* by $\mathbf{c}_1, \ldots, \mathbf{c}_g$ if, for any $i \in \{1, \ldots, t\}$, $\mathbf{c}(i) \neq 0$ implies that there exists at least a vector $\mathbf{c}' \in \{\mathbf{c}_1, \ldots, \mathbf{c}_g\}$ (depending on i), such that it is $\mathbf{c}(i) = \mathbf{c}'(i)$.

Definition 1. *Given positive integers q, k, and n, with $q \geq 2$ and $2 \leq k \leq n$, we say that a $t \times n$ matrix M with entries in $\{0, 1, \ldots, q-1\}$ is a q-ary (k, n)-strongly selective code of length t if for any k distinct columns $\mathbf{c}_{j_1}, \ldots, \mathbf{c}_{j_k}$ of M, one has that no column in $\{\mathbf{c}_{j_1}, \ldots, \mathbf{c}_{j_k}\}$ is covered by the remaining $k - 1$ columns. In other words for any $\ell \in \{1, \ldots, k\}$, there exists a row index i_ℓ such that $\mathbf{c}_{j_\ell}(i_\ell) \neq 0$ and $\mathbf{c}_{j_f}(i_\ell) \neq \mathbf{c}_{j_\ell}(i_\ell)$ for each $f \in \{1, \ldots, k\} \backslash \{\ell\}$. The minimum length of a q-ary (k, n)-strongly selective code is denoted by $t(q, k, n)$.*

For $q = 2$, the above defined codes correspond to the strongly selective families of [7] or equivalently to the $(k-1)$-cover free families of [20] and the $(k-1)$-superimposed codes of [17]. We can prove the following result.

Theorem 3. *Given positive integers q, k, and n, with $2 \leq q \leq k \leq n$, there exists a q-ary (k, n)-strongly selective code of length*

$$t \leq \frac{4(k-1)}{q-1} \left(\ln \binom{n}{k-1} + 2 \ln k + 1 \right) = O\left(\frac{k^2}{q} \log \frac{n}{k} \right).$$

We first notice that, because of the characterization of conflict resolution algorithms in terms of matrices that we have given in Sect. 1.1, Theorem 3 immediately implies the upper bound (2) of Theorem 1.

In order to prove Theorem 3 (and also Theorem 1 in its full generality), we introduce the following generalization of q-ary (k, n)-strongly selective codes.

Definition 2. *Given positive integers q, m, k, and n, with $m \leq k$, $2 \leq k \leq n$, and $q \geq 2$, we say that a $t \times n$ matrix M with entries in $\{0, 1, \ldots, q-1\}$ is a q-ary (k, m, n)-selector of size t if for any k distinct columns $\mathbf{c}_{j_1}, \ldots, \mathbf{c}_{j_k}$ of M, one has that there are at least m columns in $\{\mathbf{c}_{j_1}, \ldots, \mathbf{c}_{j_k}\}$, say $\mathbf{c}_{\ell_1}, \ldots, \mathbf{c}_{\ell_h}$, $h \geq m$, such that each column $\mathbf{c}_{\ell_r} \in \{\mathbf{c}_{\ell_1}, \ldots, \mathbf{c}_{\ell_h}\}$ is not covered by the remaining $k - 1$ columns in $\{\mathbf{c}_{j_1}, \ldots, \mathbf{c}_{j_k}\} \backslash \{\mathbf{c}_{\ell_r}\}$. The minimum size of a q-ary (k, m, n)-selector is denoted by $t_{sel}(q, k, m, n)$.*

For $q = 2$, q-ary (k, m, n)-selectors correspond to the (k, m, n)-selectors introduced in [14]. It is clear that a q-ary (k, n)-strongly selective code corresponds to a q-ary (k, k, n)-selector.

In order to prove our existential results, we need to recall the celebrated Lovász Local Lemma for the symmetric case (see [2]), as stated below.

Lemma 1. *Let $E_1, E_2, \ldots, E_b$ be events in an arbitrary probability space. Suppose that each event E_i is mutually independent of a set of all other events E_j except for at most d, and that $\Pr[E_i] \leq P$ for all $1 \leq i \leq b$. If $eP(d+1) \leq 1$, then $\Pr[\cap_{i=1}^n \bar{E}_i] > 0$, where $e = 2.71828 \ldots$ is the base of the natural logarithm.*

Using Lovász Local Lemma we will prove the following basic result.

Theorem 4. *Given positive integers q, m, k, and n, with $m \leq k$, and $2 \leq q \leq k \leq n$, there exists a q-ary (k, m, n)-selector of length*

$$t \leq \frac{4(k-1)}{(k-m+1)(q-1)} \left(\ln \binom{n}{k-1} + \ln k + \ln \binom{k}{k-m+1} + 1 \right)$$

$$= O\left(\frac{k^2}{(k-m+1)q} \log \frac{n}{k} \right).$$

Proof. Let M be a random q-ary $t \times n$ matrix such that all entries in M are chosen independently with $Pr\{M(i,j) = s\} = p = \frac{1}{k}$, for each $s \in \{1, \ldots, q\}$, and $Pr\{M(i,j) = 0\} = 1 - (q-1)p = 1 - \frac{q-1}{k}$. For a given set Υ of k columns of M, let us denote by E_Υ the event that for at least $k - m + 1$ columns, say $\mathbf{c}_{\ell_1}, \ldots, \mathbf{c}_{\ell_g} \in \Upsilon$, $g \geq k - m + 1$, one has that each column $\mathbf{c}_{\ell_r} \in \{\mathbf{c}_{\ell_1}, \ldots, \mathbf{c}_{\ell_g}\}$ is covered by the remaining $k - 1$ columns in $\Upsilon \backslash \{\mathbf{c}_{\ell_r}\}$. For a fixed subset $\Psi \subseteq \Upsilon$ of $k - m + 1$ columns, let P_Ψ denote probability that each $\mathbf{c} \in \Psi$ is covered by the columns in $\Upsilon \backslash \{\mathbf{c}\}$, i.e., for any $\mathbf{c} \in \Psi$, there exists no row index i such that $\mathbf{c}(i) \neq 0$ and $\mathbf{c}'(i) \neq \mathbf{c}(i)$ for each $\mathbf{c}' \in \Upsilon \backslash \{\mathbf{c}\}$. We have that

$$P_\Psi \leq [1 - (q-1)p(1-p)^{k-1}]^{t(k-m+1)}. \tag{6}$$

As a consequence, it holds

$$Pr\{E_\Upsilon\} \leq \binom{k}{k-m+1} [1 - (q-1)p(1-p)^{k-1}]^{t(k-m+1)}. \tag{7}$$

Observe that there are at most $k\binom{n-1}{k-1}$ k-column subsets containing one or more columns of Υ, and consequently, event E_Υ is independent from all but at most $k\binom{n-1}{k-1}$ events in $\{E_{\Upsilon'} : \Upsilon' \subseteq M, |\Upsilon'| = k\}$. Therefore, applying Lemma 1 with $P = \binom{k}{k-m+1}[1 - (q-1)p(1-p)^{k-1}]^{t(k-m+1)}$ and $d = k\binom{n-1}{t-1}$, one has that M has positive probability of being a (k, m, n)-selector if

$$e\binom{k}{k-m+1} [1 - (q-1)p(1-p)^{k-1}]^{t(k-m+1)} \left(k\binom{n-1}{k-1} + 1 \right) \leq 1.$$

By setting $p = \frac{1}{k}$ in the above inequality and requiring that

$$e\binom{k}{k-m+1} \left[1 - (q-1)\frac{1}{k}\left(1 - \frac{1}{k}\right)^{k-1} \right]^{t(k-m+1)} \left(k\binom{n-1}{k-1} + 1 \right) \leq 1, \tag{8}$$

we get that M has a strictly positive probability of being a q-ary (k, m, n)-selector if

$$t \geq \frac{\ln \left(k\binom{n-1}{k-1} + 1 \right) + \ln \binom{k}{k-m+1} + 1}{-(k-m+1)\ln \left[1 - (q-1)\frac{1}{k}\left(1 - \frac{1}{k}\right)^{k-1} \right]}. \tag{9}$$

The well known inequality

$$-\ln(1-x) > x, \quad \text{for } 0 < x < 1, \tag{10}$$

implies that the righthand side of inequality (9) is strictly smaller than

$$\frac{\ln\left(k\binom{n-1}{k-1}+1\right)+\ln\left(\frac{k}{k-m+1}\right)+1}{(k-m+1)(q-1)\frac{1}{k}\left(1-\frac{1}{k}\right)^{k-1}} \leq \frac{\ln\left(k\binom{n}{k-1}\right)+\ln\left(\frac{k}{k-m+1}\right)+1}{(k-m+1)(q-1)\frac{1}{k}\left(1-\frac{1}{k}\right)^{k-1}},$$

where the latter is a consequence of the fact that $k\binom{n-1}{k-1}+1 \leq k\binom{n}{k-1}$. Since $k \geq 2$ and $\left(1-\frac{1}{k}\right)^k$ increases with k, one has that $\left(1-\frac{1}{k}\right)^k \geq 1/4$ and it follows that the righthand side of the above inequality is at most

$$\frac{4\left(\ln\left(k\binom{n}{k-1}\right)+\ln\left(\frac{k}{k-m+1}\right)+1\right)}{(k-m+1)(q-1)\frac{1}{k}\left(1-\frac{1}{k}\right)^{-1}} = \frac{4(k-1)\left(\ln\left(k\binom{n}{k-1}\right)+\ln\left(\frac{k}{k-m+1}\right)+1\right)}{(k-m+1)(q-1)},$$

and one has that there exists a q-ary (k,m,n)-selector of length t being at most

$$\frac{4(k-1)\left(\ln\left(k\binom{n}{k-1}\right)+\ln\left(\frac{k}{k-m+1}\right)+1\right)}{(k-m+1)(q-1)}.$$

The asymptotic bound in the statement of the theorem follows from the following well known upper bound on the binomial coefficient

$$\binom{a}{b} \leq (ea/b)^b. \tag{11}$$

□

Theorem 4 implies the upper bound (1) of Theorem 1 in that a q-ary (k,m,n)-selector corresponds to a conflict resolution algorithm for the no-feedback scenario that allows all but at most $k-m$ stations to transmit with success. In order to see this, let us assume that $j \leq k$ be the number of active stations, and let M be a q-ary (k,m,n)-selector and M' be any k-column submatrix of M containing all j columns associated with the j active stations. By Definition 2, one has that M' contains m columns each of which is not covered by the remaining $k-1$ columns in M'. Therefore, at least $m-(k-j)$ of the j columns of M' associated with the j active stations are not covered by the remaining $k-1$ columns in M'. In virtue of the characterization given in Sect. 1.1, the q-ary (k,m,n)-selector corresponds to a conflict resolution algorithm for the no-feedback scenario that allows all but at most $j-(m-(k-j)) = k-m$ stations to transmit with success.

Moser and Tardos [32] gave a randomized algorithm to generate the structures whose existence is guaranteed by the Lovász Local Lemma. By exploiting their technique, we design Algorithm 1 that constructs a q-ary (k,m,n)-selector meeting the upper bound of Theorem 4. The expected number of times the

---

Algorithm 1. Algorithm for q-ary (k, m, n)-selectors.

Input: Integers t, k, m, q and n, where $q \geq 2$, $1 \leq m \leq k$, and $2 \leq k \leq n$.
Output: M : a q-ary (k, m, n)-selector.

1 Let $t := \frac{4(k-1)}{(k-m+1)(q-1)} \left(\ln \binom{n}{k-1} + \ln k + \ln \binom{k}{k-m+1} + 1 \right)$;

2 Construct a $t \times n$ matrix M where each entry $M(i, j)$ is chosen independently
  at random from $\{0, 1, \ldots, q-1\}$ with $Pr\{M(i, j) = s\} = \frac{1}{k}$, for $s \in \{1, \ldots, q\}$,
  and $Pr\{M(i, j) = 0\} = 1 - \frac{q-1}{k}$;

3 **repeat**
4   Set *flag*:= true;
5   **for** *each set C of k columns of M* **do**
6   **if** *C does not satisfy the property of Definition 2* **then**
7   Set *flag*:= false;
8   Set *missing-column-set* := C;
9   **break** ;
10   **end**
11   **end**
12   **if** *flag = false* **then**
13   Choose entries $M(i, j)$'s in the k columns of *missing-column-set*
        independently at random from $\{0, 1, \ldots, q-1\}$ with
        $Pr\{M(i, j) = s\} = \frac{1}{k}$, for $s \in \{1, \ldots, q\}$, and
14      $Pr\{M(i, j) = 0\} = 1 - \frac{q-1}{k}$;
15   **end**
16 **until** *flag = true*;
17 Output M;

---

resampling step (line 13 in Algorithm 1) is repeated is linear in n. As a consequence, for fixed k, Algorithm 1 runs in expected polynomial time. For details, see [32].

We now turn our attention to non-existential results. We prove the following theorem showing that our conflict resolution algorithm is not far from being optimal.

Theorem 5. *Given positive integers q, k, and n, with $q \geq 2$ and $2 \leq k \leq n$, the minimum length of any q-ary (k, n)-strongly selective code is*

$$t(k, n) = \Omega \left(\frac{k^2}{q \log k} \log \frac{n}{k} \right),$$ (12)

where the hidden constant is larger than or equal to $1/2$.

Proof. Let M be a $t \times n$ q-ary (k, n)-strongly selective code and let M_B be the $(q - 1) \cdot t \times n$ Boolean matrix obtained by replacing each entry by a Boolean column of length $q - 1$, with each 0-entry being replaced by the all-zero column and each non-zero entry being replaced by a Boolean column with a single entry equal to 1, according to the following mapping

$$0 \rightarrow \begin{pmatrix} 0 \\ 0 \\ \vdots \\ 0 \end{pmatrix}, \quad 1 \rightarrow \begin{pmatrix} 1 \\ 0 \\ \vdots \\ 0 \end{pmatrix}, \quad 2 \rightarrow \begin{pmatrix} 0 \\ 1 \\ \vdots \\ 0 \end{pmatrix}, \dots, \quad q-1 \rightarrow \begin{pmatrix} 0 \\ 0 \\ \vdots \\ q-1 \end{pmatrix}. \quad (13)$$

By Definition 1, one has that, for any k distinct columns $\mathbf{c}_{j_1}, \dots, \mathbf{c}_{j_k}$ of M, each column $\mathbf{c}_{j_\ell} \in \{\mathbf{c}_{j_1}, \dots, \mathbf{c}_{j_k}\}$ is such that there exists a row index i_ℓ for which $\mathbf{c}_{j_\ell}(i_\ell) = s$ for some $s \neq 0$ and $\mathbf{c}_{j_f}(i_\ell) \neq s$ for each $f \in \{1, \dots, k\} \backslash \{\ell\}$. As a consequence, entry $\mathbf{c}_{j_\ell}(i_\ell)$ is expanded into a Boolean column of M_B with all entries equal to 0 but the s-th one, whereas each entry in $\{\mathbf{c}_{j_1}(i_\ell), \dots, \mathbf{c}_{j_k}(i_\ell)\} \backslash \{\mathbf{c}_\ell(i_\ell)\}$ is expanded into a Boolean column with the s-th entries equal to 0. From the above argument it follows that M_B is a (classical) binary (k,n)-strongly selective code. The stated lower bound (12) now follows from the very well known $\Omega(\frac{k^2}{\log k} \log \frac{n}{k})$ lower bound [3,14,17,21,34] on the length of binary (k,n)-strongly selective codes. The constant hidden in that Ω-notation is larger than or equal to $1/2$ [17]. □

We now turn our attention to the construction of a combinatorial structure that will be instrumental to prove Theorem 2. We introduce the following definition.

Definition 3. *Given positive integers q, k, and n, with $q \geq 2$ and $2 \leq k \leq n$, we say that a $t \times n$ matrix M with entries in $\{0, 1, \dots, q-1\}$ is a q-ary $KG(k,n)$-code of length t if for any submatrix M' of k columns of M there exists a non-empty set of row indices $\{i_1, \dots, i_\ell\} \subseteq [t]$, with $i_1 < i_2 < \dots < i_\ell$, such that the following property holds.*

There exists a partition $\{M'_1, \dots, M'_\ell\}$ of the set of columns of M' such that for each column $\mathbf{c}$ of M'_j, one has that $\mathbf{c}(i_j) \neq 0$ and that all other columns in $M'_j, \dots, M'_\ell$ have the i_j-th entry different from $\mathbf{c}(i_j)$.

The minimum length of a q-ary $KG(k,n)$-code is denoted by $t_{KG}(q,k,n)$.

The following theorem shows that a q-ary $KG(k,n)$-code is equivalent to a scheduling protocol for our multiple-access model with feedback that allows all up to k active stations to transmit with success.

Theorem 6. *Given positive integers q, k, and n, with $q \geq 2$ and $2 \leq k \leq n$, a scheduling algorithm is a conflict resolution algorithm for a multiple-access channel $\mathcal{C}$ with feedback (comprising of $q - 1 \geq 1$ individual channels) that schedules the transmissions of n stations in such a way that all of the up to k active stations transmit successfully, if and only if the corresponding q-ary matrix is a q-ary $KG(k,n)$-code.*

Proof. The proof is omitted from this version of the paper. □

The following result, together with above Theorem 6, prove formula (4) of Theorem 2.

Theorem 7. *Given positive integers q, k, and n, with $2 \leq q \leq k \leq n$, there exists a q-ary $KG(k,n)$-code of length t with*

$$t = O\left(\frac{k}{q}\log\frac{n}{k}\right).$$

Proof. The proof is omitted from this version of the paper.   □

We now prove lower bound (5) of Theorem 2. We introduce the following definition.

Definition 4. *Given positive integers q, k, and n, with $q \geq 2$ and $2 \leq k \leq n$, we say that a $t \times n$ matrix M with entries in $\{0, 1, \ldots, q-1\}$ is a q-ary (k,n)-locally thin code of length t if for any submatrix M' of up to k columns of M there is a row index i such that, for some $s \neq 0$, one has that s occurs exactly in one entry of the i-th row of M'. The minimum length of a q-ary (k,n)-locally thin code is denoted by $t_{LT}(q,k,n)$.*

In the case $q = 2$, locally thin codes have been studied in [7,11,12,26] (each paper has essentially the same result but uses a different terminology).

Theorem 8. *Given positive integers q, k, and n, with $q \geq 2$ and $2 \leq k \leq n$, the minimum length $t_{LT}(q,k,n)$ of a q-ary (k,n)-locally thin code is*

$$t_{LT}(q,k,n) = \Omega\left(\frac{k}{q}\log\frac{n}{k}\right).$$

Proof. Let M be a $t \times n$ q-ary (k,n)-strongly selective code and let M_B be the $q \cdot t \times n$ Boolean matrix obtained, as in the proof of Theorem 5, by replacing each entry $M(i,j)$ of M by a Boolean column of length $q - 1$ according to the mapping (13) in the proof of Theorem 5. By Definition 4, one has that, for any distinct columns $\mathbf{c}_{j_1}, \ldots, \mathbf{c}_{j_p}$, $p \leq k$, of M, there exists a row index i and a column $\mathbf{c} \in \{\mathbf{c}_{j_1}, \ldots, \mathbf{c}_{j_p}\}$ such that, $\mathbf{c}(i) = s$ for some $s \neq 0$ and $\mathbf{c}_{j_f}(i) \neq s$ for each $\mathbf{c}_{j_f} \in \{\mathbf{c}_{j_1}, \ldots, \mathbf{c}_{j_p}\}\backslash\{\mathbf{c}\}$. It follows that entry $\mathbf{c}(i)$ is expanded into a Boolean column of M_B with all entries equal to 0 but the s-th one, whereas each entry in $\{\mathbf{c}_{j_1}(i), \ldots, \mathbf{c}_{j_p}(i)\}\backslash\{\mathbf{c}(i)\}$ is expanded into a Boolean column with the s-th entry equal to 0. From the above argument it follows that M_B is a binary (k,n)-locally thin code. The lower bound in the statement of the theorem is a consequence of the $\Omega\left(k \log \frac{n}{k}\right)$ lower bound of [7,11,12,26] on the length of binary (k,n)-locally thin codes.   □

Lemma 2. *Given positive integers q, k, and n, with $q \geq 2$ and $2 \leq k \leq n$, any q-ary $KG(k,n)$-code is a q-ary (k,n)-locally thin code.*

Proof. The proof is omitted from this version of the paper.   □

The following theorem is an immediate consequence of Theorem 8 and Lemma 2.

Theorem 9. *Given positive integers q, k, and n, with $q \geq 2$ and $2 \leq k \leq n$, the minimum length $t_{KG}(q, k, n)$ of a q-ary $KG(k, n)$-code is*

$$t_{KG}(q, k, n) = \Omega\left(\frac{k}{q}\log\frac{n}{k}\right).$$

Clearly, above Theorems 6 and 9 imply the lower bound (5) of Theorem 2.

5 Relationships with Frameproof Codes

The combinatorial structures of Definition 1 have similarities with the q-ary k-*frameproof codes* introduced by Boneh and Shaw [5] in the context of fingerprinting for digital data. In this section we show how our results on the minimum length of q-ary (k, n)-strongly selective codes allow to improve on those that, to our knowledge, are the best upper and lower bounds on the minimum length of q-ary k-frameproof codes for the case $q \leq k$.

A k-frameproof code has the property that for any k arbitrary distinct columns $\mathbf{c}_{j_1}, \ldots, \mathbf{c}_{j_k}$ and for any other column $\mathbf{c}$, there exists a row index i such that $\mathbf{c}(i) \notin \{\mathbf{c}_{j_1}(i), \ldots, \mathbf{c}_{j_k}(i)\}$. It is obvious that a $(k+1, n)$-strongly selective code of length t is a k-frameproof code of size n and length t. On the other hand, a k-frameproof code is not necessarily a $(k + 1, n)$-strongly selective code since the above said entry $\mathbf{c}(i)$ is possibly equal to 0. However, the minimum length of $(k+1, n)$-strongly selective codes is at most twice the minimum length of k-frameproof codes of size n. Indeed, if one is given a k-frameproof code M of size n then one can construct a $(k + 1, n)$-strongly selective code twice as long as M by concatenating the rows of M to those of the matrix obtained by replacing each entry $M(i, j)$ in M by $q - 1 - M(i, j)$. Our upper and lower bounds on the minimum length of strongly selective codes of Theorems 3 and 5, respectively, provide the following upper and lower bounds on the minimum length of k-frameproof codes of size n. We denote the minimum length of a q-ary k-frameproof code of size n by $t_{FP}(q, k, n)$.

Theorem 10. *Given positive integers q, k, and n, with $2 \leq q \leq k \leq n$, there exists a q-ary k-frameproof code of size n and length*

$$t_{FP}(q, k, n) \leq \frac{4k}{q - 1}\left(\ln\binom{n}{k} + 2\ln(k + 1) + 1\right) = O\left(\frac{k^2}{q}\log\frac{n}{k}\right).$$

We remark that the frameproof codes of Theorem 10 can be generated in pseudo-polynomial time by Algorithm 1, by replacing k with $k+1$ and setting $m = k+1$.

Theorem 11. *Given positive integers q, k, and n, with $q \geq 2$ and $2 \leq k \leq n$, the minimum length of a q-ary k-frameproof code of size n is*

$$t_{FP}(q, k, n) = \Omega\left(\frac{k^2}{q\log k}\log\frac{n}{k}\right),$$

where the hidden constant is larger than or equal to $1/4$.

In the following, we summarize those that, to our knowledge, are the best so far known bounds on the minimum length of k-frameproof codes for general values of $k \leq n$ and $q \geq 2$. In [35] it has been proved an upper bound on the size of k-frameproof codes (see Theorem 3.7 of [35]) that translates into the following lower bound on the minimum length of k-frameproof codes of size n.

$$t_{FP}(q, k, n) > k \left(\frac{\log(n/k + 1)}{\log q} - 1 \right). \tag{14}$$

We observe that our lower bound of Theorem 14 differs asymptotically from the lower bound (14) by a factor of $\frac{k \log q}{q \log k}$ which for $q < k$ is larger than 1. The authors of [37] obtained a lower bound on the maximum domain size of certain combinatorial structures, known under the name of (w_1, w_2)-separating hash families (see Corollary 19 of [37]), which translates into the following upper bound on the minimum length of k-frameproof codes.

$$t_{FP}(q, k, n) \leq \frac{k}{\log \left(\frac{q}{k}(1 + o(1)) \right)} \log \left(\frac{nk!}{k! - 1} \right). \tag{15}$$

Notice that the upper bound (15) is well defined only for $\frac{k}{q} \leq 1 + o(1)$. Theorem 10 extends the existential result to the case $q \leq k$ by providing an upper bound that differs asymptotically from the lower bound of Theorem 11 by a $\log k$ factor.

6 q-ary Strongly Selective Codes via Reed-Solomon Codes

Our upper bound of Theorem 3 and lower bound of Theorem 5 on the minimum length of q-ary (k, n)-strongly selective codes differ by a $\log k$ multiplicative factor. Clearly, the same gap transfers to our estimation of the length of optimal schedules for conflict resolution in our multichannels scenario (no feedback). In this section we show that this gap can be closed for an infinite number of the parameters. To this aim we present a construction of q-ary (k, n)-strongly selective codes based on the well known Reed-Solomon codes, whose definition we recall below.

Definition 5. *Let $\mathbb{F}_q$ be a finite field. Let $\alpha_1, \dots, \alpha_N$ be distinct elements from $\mathbb{F}_q$ and choose N and a such that $a \leq N \leq q$. We define an encoding function for Reed-Solomon code as $RS : \mathbb{F}_q^a \to F_q^N$ as follows. A message $\mathbf{m} = (m_0, m_1, \dots, m_{a-1}) \in \mathbb{F}_q^a$, with each $m_i \in \mathbb{F}_q$, is first mapped to a degree $a - 1$ polynomial $f_{\mathbf{m}}(x) = \sum_{i=0}^{a-1} m_i x^i$, $a > 2$. The encoding of $\mathbf{m}$ (i.e., the codeword associated to $\mathbf{m}$) is the evaluation of $f_{\mathbf{m}}(x)$ at all the α_i's:*

$$RS(\mathbf{m}) = (f_m(\alpha_1), f_m(\alpha_2), \dots, f_m(\alpha_N)).$$

The parameter N in the above definition is commonly taken equal to $q - 1$ and the polynomials are evaluated at all points in $\mathbb{F}_q \backslash \{0\}$. Reed-Solomon codes are linear codes with each codeword being of length N and the number of distinct codewords being equal to q^a. It is well known that Reed Solomon codes meet the Singleton bound in that they have minimum distance equal to $N - a + 1$. For undefined terms of Error Correcting Codes see [31].

The authors of [25] exploited Reed-Solomon codes to obtain *binary* (k, n)-strongly selective codes with $k = \lfloor \frac{N-1}{a-1} \rfloor$ and length $(q-1)N$. We generalize their construction to obtain (k, n)-strongly selective codes on an alphabet of arbitrary cardinality v, $2 \le v \le q$. Recall that, from basic results of finite fields theory, q can be *any* prime power [31].

Our construction works as follows. We start with a q-ary Reed-Solomon code of length $N = q - 1$ and minimum distance $N - a + 1$ and replace each entry in the codewords of this code by a column vector of length $\lfloor \frac{q-1}{v-1} \rfloor$ having a single non-zero entry. Namely, an entry equal to $x \in \{0, 1, \ldots, q-1\}$ is replaced by the $\lfloor \frac{q-1}{v-1} \rfloor$-entry column vector $\mathbf{c}_x = (\mathbf{c}_x(0), \ldots, \mathbf{c}_x(\lfloor \frac{q-1}{v-1} \rfloor))$ with

$$\mathbf{c}_x(i) = \begin{cases} x \bmod (v-1) + 1, & \text{if } i = \lfloor \frac{x}{v-1} \rfloor \\ 0 & \text{otherwise} \end{cases}$$

Let C denote the v-ary code resulting from replacing each entry $x \in \{0, 1, \ldots, q-1\}$ in the codewords of the Reed-Solomon code by $\mathbf{c}_x$. C is a code of length $t = N\lfloor \frac{q-1}{v-1} \rfloor$ in which each codeword has *exactly* N non-zero entries, given that each of the N entries in the original codeword has been replaced by a column vector with exactly one non-zero entry. We will show that any codeword in C differs from any other codeword of C in at least $a - 1$ non-zero entries. To this aim, we observe that, for any two distinct symbols $x, x' \in \{0, 1, \ldots, q-1\}$, we have that at least one of the following holds: $x \bmod (v-1) + 1 \ne x' \bmod (v-1) + 1$, $\lfloor \frac{x}{v-1} \rfloor \ne \lfloor \frac{x'}{v-1} \rfloor$. Consequently, $\mathbf{c}_x$ and $\mathbf{c}_{x'}$ differ in at least one entry. More precisely, there exists either an index i such that $\mathbf{c}_x(i) \ne 0$, $\mathbf{c}_{x'}(i) \ne 0$ and $\mathbf{c}_x(i) \ne \mathbf{c}_{x'}(i)$, or there exists two indices i and i' such that $0 = \mathbf{c}_x(i) \ne \mathbf{c}_{x'}(i)$ and $\mathbf{c}_x(i') \ne \mathbf{c}_{x'}(i') = 0$. In both cases, $\mathbf{c}_x$ has a non-zero entry which is different from the corresponding entry of $\mathbf{c}_{x'}$, and $\mathbf{c}_{x'}$ has a non-zero entry which is different from the corresponding entry of $\mathbf{c}_x$. It follows that if two codewords of the Reed-Solomon code differ in d entries then each of the corresponding codewords in C has d non-zero entries at which it differs from the other codeword. For any pair of codewords in C, the number of these entries is at least as large as the minimum distance of the Reed-Solomon code, namely $N - a + 1$. This, along with the fact that each codeword in C has N non zero entries, implies that for any two codewords $\mathbf{c}_1, \mathbf{c}_2 \in C$ there are at most $N - (N - a + 1) = a - 1$ indices i's such that $\mathbf{c}_1(i) = \mathbf{c}_2(i) \ne 0$. As a consequence, for any codeword $\mathbf{c} \in C$ and for any other k codewords $\mathbf{c}_1, \ldots, \mathbf{c}_k \in C$, $\mathbf{c}$ has at most $a - 1$ non-zero entries in common with each of $\mathbf{c}_1, \ldots, \mathbf{c}_k$, and therefore, there exist at most $k(a - 1)$ indices i's such that $\mathbf{c}(i) \ne 0$ and $\mathbf{c}_j(i) = \mathbf{c}(i)$ for some $j \in \{1, \ldots, k\}$. If we choose $k = \left\lfloor \frac{N-1}{a-1} \right\rfloor = \left\lceil \frac{N}{a-1} \right\rceil - 1$ then it holds $k(a - 1) < N$ and, by the above argument, there is an index i such that $\mathbf{c}(i) \ne 0$ and $\mathbf{c}_1(i) \ne \mathbf{c}(i), \mathbf{c}_2(i) \ne \mathbf{c}(i), \ldots, \mathbf{c}_k(i) \ne \mathbf{c}(i)$. In other words, C is a v-ary (k, n)-strongly selective code, where $n = q^a$, and the length t and the "selection" capability k are equal to

$$t = N \left\lfloor \frac{q-1}{v-1} \right\rfloor \qquad k = \left\lceil \frac{N}{a-1} \right\rceil - 1. \tag{16}$$

Now, consider the parameter a constant, and let instead the size $q = N + 1$ of the field F_q arbitrarily vary among the set of prime powers (there are many). Moreover, we also constraint the value of k in such a way that $k \leq bn^{1/a}$, for some constant b. The relations between t, n, k and the alphabet cardinality of our v-ary (k, n)-strongly selective code can be thus so evaluated

$$t \leq \frac{(q-1)^2}{v-1} = \frac{1}{v-1} \frac{(q-1)^2}{(a-1)^2} (a-1)^2 \qquad \text{(from the first identity in (16))}$$

$$\leq (a-1)^2 \frac{(k+1)^2}{v-1} \leq (a-1)a \frac{(k+1)^2}{v-1} \qquad \text{(from the second identity in (16))}$$

$$\leq (a-1) \frac{(k+1)^2}{(v-1)(\log k - \log b)} \log n \qquad \text{(because we are in the regime } k \leq bn^{1/a}).$$

Since a and b are constant, we have found an infinity of values of the parameters n and k for which there exist v-ary (k, n)-strongly selective codes of length t matching (asymptotically) the lower bound (12). It has not escaped our attention that, *under* the constraints outlined above, our result also implies that it is possible to construct in *polynomial* time, binary (k, n)-strongly selective codes of *optimal* length $\Theta\left(\frac{k^2}{\log k} \log n\right)$. Recently, an unconstrained tight bound $\Theta\left(\frac{k^2}{\log k} \log n\right)$ was claimed in [18], however this claim has now been retracted in [19].

References

1. I.802.11 Wireless LAN MAC and Physical Layer Specification. http://www.ieee802.org/
2. Alon, N., Spencer, J.H.: The probabilistic method. In: Wiley-Interscience Series in Discrete Mathematics and Optimization, 3rd edn. Wiley, Hoboken (2008)
3. Alon, N., Asodi, V.: Learning a hidden subgraph. SIAM J. Discrete Math. **18**, 697–712 (2005)
4. Avgouleas, I., Angelakis, V., Pappas, N.: Utilizing multiple full-duplex relays in wireless systems with multiple packet reception. In: 19th IEEE International Workshop on Computer Aided Modeling and Design of Communication Links and Networks (CAMAD), pp. 193–197 (2014)
5. Boneh, D., Shaw, J.: Collusion-free fingerprinting for digital data. IEEE Trans. Inf. Theory **44**, 1897–1905 (1998)
6. Bluetooth Consortium: Bluetooth Specification. https://www.bluetooth.com/
7. Clementi, A.E.F., Monti, A., Silvestri, R.: Distributed broadcast in radio networks of unknown topology. Theor. Comput. Sci. **302**(1–3), 337–364 (2003)
8. Chlamtac, I., Faragó, A.: An optimal channel access protocol with multiple reception capacity. IEEE Trans. Comput. **43**, 480–484 (1994)

9. Chlebus, B.S.: Randomized communication in radio networks. In: Pardalos, P.M., Rajasekaran, S., Reif, J.H., Roli, J.D.P. (eds.) Handbook on Randomized Computing, vol. 1, pp. 401–456. Kluwer Academic Publishers (2001)

10. Chrobak, M., Gąsieniec, L., Rytter, W.: Fast broadcasting and fast broadcasting and gossiping in radio networks. J. Algorithms **43**(2), 177–189 (2002)

11. Cohen, G.D.: Applications of coding theory to communication combinatorial problems. Discrete Math. **83**(2–3), 237–248 (1990)

12. Csűrös, M., Ruszinkó, M.: Single-user tracing and disjointly superimposed codes. IEEE Trans. Inf. Theory **51**(4), 1606–1611 (2005)

13. Daum, S., Kuhn, F., Newport, C.: Efficient symmetry breaking in multi-channel radio networks. DISC **2012**, 238–252 (2012)

14. De Bonis, A., Gąsieniec, L., Vaccaro, U.: Optimal two-stage algorithms for group testing problems. SIAM J. Comput. **34**(5), 1253–1270 (2005)

15. Dolev, S., Gilbert, S., Khabbazian, M., Newport, C.: Leveraging channel diversity to gain efficiency and robustness for wireless broadcast. In: Peleg, D. (ed.) DISC 2011. LNCS, vol. 6950, pp. 252–267. Springer, Berlin (2011). doi:10.1007/978-3-642-24100-0_25

16. Du, D.Z., Hwang, F.K.: Combinatorial Group Testing and Its Applications. World Scientific, River Edge (2000)

17. D'yachkov, A.G., Rykov, V.V.: Bounds on the length of disjunct codes. Problemy Peredachi Informatsii **18**(3), 7–13 (1982)

18. D'yachkov, A.G., Vorobév, I.V., Polyansky, N.A., Yu, V.: Shchukin: bounds on the rate of disjunctive codes. Probl. Inf. Transm. **50**(1), 27–56 (2014)

19. D'yachkov, A.G., Vorobév, I.V., Polyansky, N.A., Yu, V.: Shchukin: erratum to: bounds on the rate of disjunctive codes. Probl. Inf. Transm. **50**, 27 (2014). Problems of Information Transmission, **52**(2), 200 (2016)

20. Erdös, P., Frankl, P., Füredi, Z.: Families of finite sets in which no set is covered by the union of r others. Israel J. Math. **51**, 75–89 (1985)

21. Füredi, Z.: On r-cover-free families. J. Comb. Theory Ser. A **73**, 172–173 (1996)

22. Gyory, S.: Coding for a multiple access OR channel: a survey. Discrete Appl. Math. **156**, 1407–1430 (2008)

23. Halldórsson, M., Wang, Y., Yu, D.: Leveraging multiple channels in ad hoc networks. In: PODC 2015, pp. 431–440 (2015)

24. Holzer, S., Pignolet, Y., Smula, J., Wattenhofer, R.: Time-optimal information exchange on multiple channels. In: FOMC 2011, pp. 69–76 (2011)

25. Kautz, W.H., Singleton, R.C.: Nonrandom binary superimposed codes. IEEE Trans Inf. Theory **10**, 363–377 (1964)

26. Khasin, L.S.: Conflict resolution in a multiple access channel. Probl. Peredachi Inf. **25**(4), 63–68 (1989)

27. Komlós, J., Greenberg, A.G.: An asymptotically fast non-adaptive algorithm for conflict resolution in multiple-access channels. IEEE Trans. Inf. Theory **31**(2), 302–306 (1985)

28. Kowalski, D.R.: On selection problem in radio networks. In: PODC 2005, pp. 158–166. ACM Press (2005)

29. Krishnaswamy, H., Zussman, G.: 1 chip, 2× the bandwidth. IEEE Spectr. **53**(7), 38–54 (2016)

30. Massey, J.L., Mathys, P.: The collision channel without feedback. IEEE Trans. Inf. Theory **31**, 192–204 (1985)

31. MacWilliams, F.J., Sloane, N.J.A.: The Theory of Error-Correcting Codes. North Holland Publishing Co., Amsterdam (1977)

32. Moser, R.A., Tardos, G.: A constructive proof of the general Lovász local lemma. J. ACM **57**, 11–15 (2010)
33. Pappas, N., Kountouris, M., Ephremides, A., Traganitis, A.: Relay-assisted multiple access with full-duplex multi-packet reception. IEEE Trans. Wirel. Commun. **14**, 3544–3558 (2015)
34. Ruszinkó, M.: On the upper bound of the size of the r-cover-free families. J. Comb. Theory Ser. A **66**, 302–310 (1994)
35. Staddon, J.N., Stinson, D.R., Wei, R.: Combinatorial properties of frameproof and traceability codes. IEEE Trans. Inf. Theory **47**, 1042–1049 (2001)
36. Stinson, D.R., Wei, R., Chen, K.: On generalized separating hash families. J. Comb. Theory Ser. A **115**, 105–120 (2008)
37. Stinson, D.R., Zaverucha, G.M.: Some improved bounds for secure frameproof codes and related separating hash families. IEEE Trans. Inf. Theory **54**, 2508–2514 (2008)
38. Stinson, D.R., Wei, R., Zhu, L.: New constructions for perfect hash families and related structures using combinatorial designs and codes. J. Comb. Des. **8**, 189–200 (2000)
39. Yan, Y., Yu, D., Wang, Y., Yu, J., Lau, F.C.: Bounded information dissemination in multi-channel wireless networks. J. Comb. Optim. **31**, 996–1012 (2016)
40. Yu, D., Wang, Y., Yan, Y., Yu, J., Lau, F.C.: Speedup of information exchange using multiple channels in wireless ad hoc networks. In: 2015 IEEE Conference on Computer Communication (INFOCOM), pp. 2029–20137 (2015)
41. Wang, Y., Wang, Y., Yu, D., Yu, J., Lau, F.C.: Information exchange with collision detection on multiple channels. J. Comb. Optim. **31**, 118–135 (2016)

The Impact of the Gabriel Subgraph of the Visibility Graph on the Gathering of Mobile Autonomous Robots

Shouwei Li, Friedhelm Meyer auf der Heide, and Pavel Podlipyan[(✉)]

Heinz Nixdorf Institute and Department of Computer Science,
Paderborn University, Fürstenallee 11, 33102 Paderborn, Germany
{shouwei.li,fmadh,pavel.podlipyan}@upb.de

Abstract. In this work, we reconsider the well-known Go-To-The-Center algorithm due to Ando, Suzuki, and Yamashita [2] for gathering in the plane n autonomous mobile robots with limited viewing range. The above authors have introduced it as a discrete, round-based algorithm and proved its correctness. In [8], by Degener et al. it is shown that the algorithm gathers in $\Theta\left(n^2\right)$ rounds. Remarkably, this algorithm exploits the fact, that during its execution, many collisions of robots occur. Such collisions are interpreted as a success because it is assumed that such collided robots behave the same from now on. This is o.k. under the assumption, those robots have no extent. Otherwise, collisions should be avoided.

In this paper, we consider a continuous Go-To-The-Center (GTC) strategy in which the robots continuously observe the positions of their neighbors and adapt their speed (assuming a speed limit) and direction. Our first results are time bounds of $O\left(n^2\right)$ for gathering in two-dimensional Euclidean space, and $\Theta\left(n\right)$ for the one-dimensional case.

Our main contribution is the introduction and evaluation of a continuous algorithm which performs Go-To-The-Center considering only the neighbors of a robots w.r.t. the Gabriel subgraph of the visibility graph (GTGC). We show that this modification still correctly executes gathering in one and two dimensions, with the same time bounds as above. Simulations exhibit a severe difference of the behavior of the GTC and the GTGC strategy: Whereas lots of collisions occur during a run of the GTC strategy, typically only one, namely the final collision occurs during a run of the GTGC strategy. We can prove this "collisionless property" of the GTGC algorithm for the one-dimensional case. In the case of the two-dimensional Euclidean space, we conjecture that the "collisionless property" holds for almost every initial configuration.

This work was partially supported by the German Research Foundation (DFG) within the Collaborative Research Center "On-The-Fly Computing" (SFB 901) and the International Graduate School "Dynamic Intelligent Systems".

This work is submitted to Distributed & Mobile track of ALGOSENSORS 2016.

M. Chrobak et al. (Eds.): ALGOSENSORS 2016, LNCS 10050, pp. 62–79, 2017.
DOI: 10.1007/978-3-319-53058-1_5

1   Introduction

In this paper, we study the gathering problem. A group of n autonomous mobile robots is needed to be gathered at a common, not predefined point. The movement of the robots solely depends on the relative positions of other robots in their viewing range. There is no global view, communication or long term memory.

There are two standard time and activation models in use, namely discrete and continuous. The asynchronous discrete time model is presented by Cohen and Peleg in [4]. Ando, Suzuki, and Yamashita in [2] show that robots with limited visibility acting in synchronous discrete time model gather at one not predefined point using Go-To-The-Center (GTC) algorithm. Recently it was shown by Degener et al. in [8] that GTC algorithm, described in [2] gathers the group of robots in $\Theta\left(n^2\right)$ rounds. The correctness of GTC strategy w.r.t. different proximity graphs in arbitrary dimensions were presented by Cortes et al. in [6]. The Gathering problem with continuous time model for the first time was presented by Gordon, Wagner, and Bruckstein in [10] and then analyzed by Kempkes, Kling and Meyer auf der Heide in [12]. Unlike in conventional discrete time models, robots do not act in rounds, but rather continuously adjust directions of movement towards calculated target points, while moving with constant velocity. As a result, runtime could not be defined as the number of rounds in discrete models. Instead, it is defined in [12] as the time that robots need to gather at one point.

Nevertheless, independent of time and activation model, gathering strategies either ignore or exploit merge (collision) of the robots. If merges are not ignored, then they are accounted as positive events that represent the progress of the gathering process. However, in practice collisions between moving objects are usually avoided. The need to avoid collisions becomes clearer when robots gain extent. Robots with an extent were considered by Czyzowicz et al. in [7] for the first time. In this work, authors presented the solution for three and four robots by examining all possible cases exhaustively. In order to deal with the prohibition of mergers caused by extent the authors of [7] redefine the gathering itself. Since robots with an extent are not allowed to occupy the same position, gathering in [7] means forming a configuration for which the union of all discs representing robots is connected. For the same model as in [7] a general solution for a number of robots greater than four is presented by Agathangelou et al. in [1]. Compared to [7], the additional assumption in [1] is chirality. The gathering of robots with an extent is defined in a different way by Cord-Landwehr et al. in [5]. The goal of the robots is to gather without touching, in such a way that discs representing robots do not intersect. Robots gather around a predefined point that is already known to every robot. Gathering in the discrete time model proposed in [5] is proved to be done in $O\left(nR\right)$ rounds, where n is the number of robots and R is the distance from the gathering point to the farthest robot. Unlike in [7] and [1], robots in [5] have limited vision. The robots with limited vision but without an extend are considered by Pagli, Prencipe, and Viglietta in [14]. This work considers the Near Gathering problem, where the aim of the robots is to get close enough to be in vision range of each other without collisions and switch in this way to the setting where algorithms that utilize global vision will work. Furthermore, robots in [14] are equipped with a compass so that they have common local coordinate systems.

Inspired by the results mentioned above we would like to continue this line of research and focus on gathering without collisions in the continuous time model. In this paper we consider the continuous variant of GTC strategy introduced in [2] and based on it, we present GTGC algorithm. We use the Gabriel graph presented by Gabriel and Sokal in [9] to modify the original GTC algorithm. The modified GTGC algorithm considers only the neighbors of a robots w.r.t. the Gabriel subgraph of the visibility graph.

Our first result states that in one-dimensional Euclidean space GTC and GTGC algorithms gather groups of n robots in time $\Theta(n)$. In two dimensions GTC and GTGC algorithms gather groups of n robots in time $O(n^2)$. Besides the runtime analysis, the part of our main result is the evaluation of emergent properties of GTGC algorithm. The modification of original GTC algorithm has a severe impact on the behavior of robots. We show that in one dimension no collisions appear for every initial configuration if robots use GTGC algorithm. The properties of the gathering process where robots use GTGC algorithms in two dimensions we evaluate via experiments. Experimental results strongly suggest that if robots use the GTGC algorithm in 2D, then there are no collisions among the robots during the gathering process for almost every initial configuration, except for the final step, when all robots simultaneously meet at one point.

2   Problem Description

We consider a set $R = \{r_1, \ldots, r_n\}$ of n autonomous mobile robots with limited viewing range. The position of the robot r_i in space and time we denote by $r_i(t) \in X$, where $X \subset \mathbb{R}^d \times \mathbb{R}^+$. We are going to consider robots in Euclidean space of dimension 1 and 2.

The Euclidean distance between two robots r_i and r_j is represented by $|r_i(t), r_j(t)|$. Robots agree on the unit distance. Two robots are neighbors, i.e. see each other, if $|r_i(t), r_j(t)| \leq 1$, where 1 is the viewing range.

The set of robots that consists of the robot r_i itself and all its neighbors at time t is called the unit disc graph neighborhood of r_i and denoted by $UDG_t(r_i)$. The unit disc graph (UDG) defined on all robots at some point in time t is denoted by $UDG(R) = (R, E)$, where $(r_i, r_j) \in E$ iff $|r_i(t), r_j(t)| \leq 1$. The disposition of robots at some point in time t on the plane is called *configuration*. In order to simplify the notation, we skip the time in the designation of unit disc neighborhood and unit disc graph, unless time is needed to be mentioned explicitly.

The disposition of robots at time 0 is called *initial configuration*. Initial configuration is arbitrary except that the unit disc graph $UDG(R)$ over all robots is connected and all robots have distinct positions. The goal is to gather all robots in one not predefined point.

Robots are oblivious, act depending only on the information about the current point of time. Robots are anonymous, i.e., they do not have IDs. They are silent which means that they do not communicate, except they can observe the position of their neighbors. The accuracy of the robots is assumed to be perfect; namely, measurements (e.g. relative position of its neighbors) performed by the robot are exact as well as adjustments to the speed and the direction of the robot are exact and not delayed compared to the measurement.

2.1   Time Model

We consider robots in the continuous time model, which was for the first time proposed by Gordon et al. in [10] and examined after by Kempkes et al. in [12]. Instead of acting in rounds with respect to the discrete time model, robots continuously measure the relative positions of their neighbors and adjust their own velocity and direction of movement. The velocity of the robot solely depends on relative positions of neighboring robots at the current point of time. The velocity of the robot may change in a non-continuous way since robots measure relative positions of their neighbors without delay and instantly adjusts own movement with respect to the measurements. The speed of the robot is assumed to be at most 1.

The authors of [12] specify that the continuous time model may be viewed as the extreme instance of the discrete classical Look-Compute-Move (LCM) model. Assuming a speed limit of one, the continuous time model arises from the discrete LCM model by fixing the maximum distance traveled per round by δ and letting $\delta \to 0$.

2.2   The Algorithm

In this paper, we reconsider the well-known Go-To-The-Center algorithm due to Ando, Suzuki, and Yamashita [2]. The smallest or minimum enclosing circle (MEC) plays a central role in this algorithm. *Minimum enclosing circle* is the smallest circle that contains all of a given set of points in the Euclidean space. Let us consider first two-dimensional case. For the MEC in 2D, George Chrystal has shown the following properties.

Proposition 1. (Chrystal [3]) *Let C be minimum enclosing circle of a point set S. Then either:*

1. *there are two points $m_1, m_2 \in S$ on the circumference of C such that the line segment $(m_1 m_2)$ is diameter of C, or*
2. *there are three points $m_1, m_2, m_3 \in S$ such that C circumscribes $\triangle m_1 m_2 m_3$ and center c of C is inside $\triangle m_1 m_2 m_3$, which means that $\triangle m_1 m_2 m_3$ is acute or at most right triangle.*

Let us consider a robot $r \in R$ at some point in time t together with its unit disc graph neighborhood $UDG(r)$. The minimum enclosing circle $C(r)$ encircles all unit disc neighbors in $UDG(r)$.

With respect to the cases considered in Proposition 1 we are going to call *minimum enclosing set* of the robot r (at time t) the following sets $MEC(r) =$

$\{m_1, m_2\}$ or $MEC(r) = \{m_1, m_2, m_3\}$. We say that robots in sets $\{m_1, m_2\}$ or $\{m_1, m_2, m_3\}$ *form* the minimum enclosing circle $C(r)$ of robot r. Note that robot r might belong to its own minimum enclosing set, e.g. $MEC(r) = \{m_1, r\}$. In one dimensional case the smallest enclosing circle that encircles more than one robot is always formed by two robots.

Minimum enclosing circle of a point set is unique [3]. The minimum enclosing circle (sphere) can be found in linear time in Euclidean space of any constant dimension [13]. In case there is more than one minimum enclosing set $MEC(r)$ that may form C, then we assume that robot selects one arbitrary suitable set. We skip the time in the notation of minimum enclosing set and circle unless time is needed to be mentioned explicitly.

The next important structure we need for our algorithm is the Gabriel graph. It is introduced into two-dimensional Euclidean space by Gabriel and Sokal in [9] as follows:

Definition 1 (Gabriel graph criterion). *Any two robots u, v are connected if no other robot w is present within the circle whose diameter is line segment uw. We are going to call this circle Gabriel circle.*

We are going to denote by $GG(r)$ (at time t) the subgraph obtained from the unit disc graph neighborhood $UDG(r)$ by applying to it the Gabriel graph criterion. We are going to call $GG(r)$ the unit *Gabriel graph neighborhood* of robot r. The Gabriel graph defined on all robots is denoted by $GG(R)$. We skip time in the notation of Gabriel graph and Gabriel graph neighborhood, unless it is needed to be mentioned explicitly.

In the one-dimensional case, the Gabriel graph criterion is reduced to the following statement. Any two robots u, v are connected if, and only if, all other robots are outside the line segment uv.

Now we are going to introduce the modified version of the original GTC algorithm. Our modification is going to be called Go-To-The-Gabriel-Center (GTGC) algorithm. At each point of time t, each robot r does the following:

Algorithm 1. GTGC

Require: Initial configuration
 1: Robot r observes the positions of its neighbors in its $GG(r)$.
 2: Robot r computes the minimum circle $C(r)$ enclosing $GG(r)$. The center $T(r)$ of
    $C(r)$ is the *target point* of r.
 3: Robot r moves:
 4: **if** r is already at $T(r)$ **then**
 5:     Robot r remains at $T(r)$ and moves in the same way as the target point does.
 6: **else**
 7:     Robot r moves with maximum speed 1 towards $T(r)$.

If we are going to calculate MEC of $UDG(r)$ for robot r in Algorithm 1, then we will have exactly GTC algorithm from [2] for the continuous time model, except in continuous time model we do not need to calculate *LIMIT* circles as in [2] in order to keep connectivity.

3 Correctness and Runtime Analysis

First, in this section we are going to show in Lemmas 1 and 2 that both algorithms preserve connectivity of underlying graphs. Then we are going to analyze their runtime.

Lemma 1. *Let us consider a group of robots R on the Euclidean plane that follows GTC algorithm. If $\{u, w\}$ is an edge in $UDG(R)$ at time t_0, then $\{u, w\}$ is an edge in $UDG(R)$ at $\forall t_1 \geq t_0$.*

Proof. Let us consider robots u and one of unit disc edges of this robot, namely $\{u, w\}, w \in UDG(u)$. The area Q_u is an intersection of unit discs of all unit disc neighbors $UDG(u)$ at time t_0. The center $T(u)$ of the minimum enclosing circle $C(u)$ of $UDG(u)$ at time t_0 is situated inside of area Q_u as well, since radius of $C(u)$ is at most radius of unit disc and $C(u)$ encircles all robots in $UDG(u)$.

Robot u, that executes GTC algorithm, goes towards its target point, namely the center of minimum enclosing circle $T(u)$. Line segment between $T(u)$ and u is entirely contained in Q_u. Assume that at the end of the time interval $[t_0, t_1]$ the distance between robots u and w is greater than 1. Since the motion of the robots that follow GTC algorithm is continuous, there exists point in time $t' \in [t_0, t_1]$, such that at that time t' the distance between u and w is exactly one.

Let L be an intersection of unit discs of the robot u and w at time t'. Convex areas Q_u and Q_w are situated inside of L, thus at time t' robot u (as well as w) can move only inside of L. This contradicts our assumption, thus if $\{u, w\}$ is an edge in $UDG(R)$ at time t_0. Then $\{u, w\}$ is an edge in $UDG(R)$ at $\forall t_1 \geq t_0$. □

Lemma 1 allows us to conclude that if unit disc graph is connected at time t, then robots that use GTC will stay connected at any $t' \geq t$. Connectivity preservation of GTC in one-dimensional follows from Lemma 1 as well. Next, we consider GTGC.

Lemma 2. *Let us consider group of robots R on Euclidean plane that follows GTGC. If $\{u, w\}$ is an edge in $GG(R)$ at time t_0. Then $\{u, w\}$ is an edge in $GG(R)$ at $\forall t_1 \geq t_0$ or there is a path from u to w in $GG(R), \forall t_1 \geq t_0$.*

Proof. First of all, we shall note that every edge in unit Gabriel graph is an edge of unit disc graph. Because of that if an edge $\{u, w\}$) does not violate Gabriel graph criterion during the time interval $[t_0, t_1]$ it remains connected as it is shown in Lemma 1.

If an edge $\{u, w\}$ is not an edge in $GG(R)$ at time t_1 due to Gabriel graph criterion. Then there is at least one robot w that causes connection in $GG(R)$ between u and w to be removed at some point in time $t' \in [t_0, t_1]$. Therefore, there shall be an alternative path via robot that causes connection to be removed [11]. That is why there is a path from u to w in $GG(R), \forall t_1 \geq t_0$. □

Lemma 2 allows us to conclude that if unit Gabriel graph is connected at time t, then robots that use GTGC will stay connected at any $t' \geq t$. Connectivity property of GTGC in one-dimensional follows also from Lemma 2.

Let us now prepare for the runtime analysis and define the progress measure for our algorithms. We consider the global convex hull $H(R)$ around the positions of all robots at time t. We are particularly interested in robots that are corners of a global convex hull. Namely we consider the set of robots $CH(R) = \{c_i \in H(R) : \alpha_i(t) \in [0, \pi), i \in [1, k], k \leq n\}$, where n is total number of robots, k is number of robots that belong to the convex hull and have internal angle $\alpha_i(t) \in [0, \pi)$. The set $CH(R)$ we are going to call global *corner hull.*

Let $l(t)$ be the length of the global corner hull $CH(R)$. Let $l'(t)$ be the speed with which the length of the global corner hull is changing. The length of the global corner hull we are going to use a progress measure. This measure fits for the analysis of both GTC and GTGC algorithm. The upcoming Lemmas 3 and 4 hold for both algorithms as well. Lemma 3 states that the length $l(t)$ of the corner hull $CH(R)$ is a monotonically decreasing function, and Lemma 4 states that the monotonicity is strict. We are going to abuse the notation and use bar, i.e. $\overline{CH}(R)$ to refer to the closed areas defined by the sets under the bar. For example $v \in \overline{CH}(R)$ would refer to the position associated with the robot v and would mean that v is positioned inside of global corner hull. On the other hand $u \in CH(R)$ refers to the robots itself and means that u belongs to global corner hull.

Lemma 3. *The target point $T(r)$ of any robot $r \in R$ is inside $\overline{CH}(R)$.*

Proof.   Let's consider minimum enclosing set $MEC(r)$ of robot r. All robots of $MEC(r)$ including r itself are inside of $\overline{CH}(R)$.

In case $MEC(r)$ consists of two robots m_1 and m_2, the target point is midpoint of the line segment $m_1 m_2$. Since $\overline{CH}(R)$ is convex and m_1 and m_2 are in it, line segment between m_1 and m_2 is inside $\overline{CH}(R)$ too. Therefore, the target point is in $\overline{CH}(R)$.

For the case where $MEC(r)$ consists of three robots m_1, m_2 and m_3, the target point is the center of circumscribed circle around $\triangle m_1 m_2 m_3$. Let us consider without loss of generality one of the robots in $MEC(r)$, e.g. m_1. Let us now draw the line l from m_1 through the target point $z(r)$. It is clear that the line l intersects the opposite side of the $\triangle m_1 m_2 m_3$, namely $m_2 m_3$ at some point a. Since $\overline{CH}(R)$ is convex and m_2 and m_3 are in it, line segment between m_2 and m_3 is inside $\overline{CH}(R)$ too, as well as point a. The target point $z(r)$ must be also be inside $\overline{CH}(R)$ because the same argument applies for the line segment $m_1 a$.   □

Lemma 4. *If robot $c_i \in CH(R)$ then the target point $T(c_i)$ does not coincide with the position of the robot $c_i(t)$.*

The proof of Lemma 4 is easily derived from the following proposition.

Proposition 2. (Chrystal [3]) *Let C be the minimum enclosing circle of a set of $n \geq 2$ points. Then there is no point-free arc with length greater than π.*

If for some robot $c_i \in CH(R)$ the center of the minimum enclosing circle, i.e. target point $T(r)$ coincides with the position of robot r, then there is a robot-free arc (the part of circle outside of $CH(R)$) with a length greater than π. This contradicts Proposition 2.

By Lemma 4 the length of the corner hull $l(t)$ is strictly monotonically decreasing function since it is guaranteed that all robots that belong to the $CH(R)$ will move inside of the hull $CH(R)$ with speed 1. Note that Lemmas 3 and 4 do not depend on the changes in GG(R).

Now we have enough information to analyze the runtime.

Theorem 1. *In one-dimensional Euclidean space, the GTC algorithm and the GTGC algorithm gather n robots in time $\Theta(n)$.*

Proof. The corner hull $CH(R)$ around the configuration where all robots are placed on the same line consists only of two robots, let us call them c_1 and c_2. Due to Lemma 4, the center of minimum enclosing circle can not coincide with position of the robot $c_1(t) \in CH(R)$, thus robot c_1 moves. By Lemma 3, the velocity vector $v_1(t)$ of robot c_1 points inside of the corner hull. The same holds for c_2. Robots c_1 and c_2 move with speed 1, thus the speed $l'(t)$ with which the length of the global corner hull $l(t)$ decreases is 4. The corner hull consists of two line segments c_1c_2 and c_2c_1 both decrease with speed 2.

At most the length of the corner hull $l(t)$ in one dimension is $2(n-1)$, thus after the time

$$t_* = \frac{l(t)}{4} \leq \frac{(n-1)}{2} \tag{1}$$

all robots are gathered at one point in one dimensional case. In the case where all robots are placed at th maximum distance (viewing range) apart from each other our inequality 1 becomes an equality. □

Next, we are going to analyze the runtime of GTC and GTGC in the two-dimensional case. The upcoming Lemmas 5 and 6 hold for both algorithms.

Lemma 5. *At any moment of time t length of corner hull $l(t)$ decreases with speed $l'(t) \geq 8/n$.*

Proof. Lemma 4 states that the target point $T(c_i)$ can not coincide with position of the robot $c_i(t) \in CH(R)$, thus robot c_i moves. As well by Lemma 3, the velocity vector $v_i(t)$ of robot c_i points inside of the corner hull. We do not know how $v_i(t)$ divides internal angle $\alpha_i(t)$. It is illustrated on Fig. 1. Due to this we introduce the parameter $\vartheta_i(t) \in [0,1]$ that defines angles between velocity vector $v_i(t)$ and edges of the corner hull adjacent to robot c_i. Using parameter $\vartheta_i(t)$, the internal angle $\alpha_i(t)$ is given by

$$\alpha_i(t) = \alpha_i(t)\vartheta_i(t) + \alpha_i(t)(1 - \vartheta_i(t)). \tag{2}$$

Note that a change of $UDG(c_i)$ at time t as well as $GG(c_i)$ affects just the direction of $v_i(t)$ and causes a change of parameter $\vartheta_i(t) \in [0,1]$ within a given range.

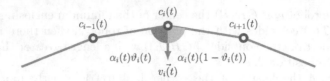

Fig. 1. Velocity $v_i(t)$ of robot c_i.

Now we are going to consider an edge of corner hull between robots c_i and c_{i+1}. The length of this edge is given by $d_{i,i+1}(t) = |c_i(t), c_{i+1}(t)|$. The velocity $d'_{i,i+1}(t)$, with which the length is changing is represented by $d'_{i,i+1}(t) = Pv_i(t) + Pv_{i+1}(t)$, where $Pv_i(t)$, $Pv_{i+1}(t)$ are orthogonal projection of velocity of robots c_i and c_{i+1} onto the line between this robots. Projections, as well as $d'_{i,i+1}(t)$ could be represented by scalar velocities $d'_{i,i+1}(t), Pv_i(t), Pv_{i+1}(t)$ where sign of the scalar will represent whether corresponding component of speed increases or decreases length $d_{i,i+1}(t)$. Further, in order to simplify the notation we are going to skip time t. The value of Pv_i depends on $\alpha_i(1 - \vartheta_i)$ and could be represented as follows

$$Pv_i(\alpha_i(1 - \vartheta_i)) = \begin{cases} \cos(\alpha_i(1 - \vartheta_i)) & \text{if } \alpha_i(1 - \vartheta_i) \leq \pi/2; \\ -\cos(\alpha_i(1 - \vartheta_i)) & \text{if } \alpha_i(1 - \vartheta_i) > \pi/2. \end{cases} \tag{3}$$

In the same way we represent Pv_{i+1} that depends on $\alpha_{i+1}\vartheta_{i+1}$. The scalar velocity with which the length of the edge between c_i and c_{i+1} is changing now can be expressed as follows

$$d'_{i,i+1}(\alpha_i(1 - \vartheta_i), \alpha_{i+1}\vartheta_{i+1}) = Pv_i(\alpha_i(1 - \vartheta_i)) + Pv_{i+1}(\alpha_{i+1}\vartheta_{i+1}). \tag{4}$$

Overall speed with witch the length of the corner hull is changing at time t could be represented by the sum of scalar velocities of each edge of corner hull

$$l' = \sum_{i=1}^{k} d'_{i,i+1}(\alpha_i(1 - \vartheta_i), \alpha_{i+1}\vartheta_{i+1}) = \sum_{i=1}^{k} (Pv_i(\alpha_i(1 - \vartheta_i)) + Pv_{i+1}(\alpha_{i+1}\vartheta_{i+1})). \tag{5}$$

In this sum the element $d'_{k,k+1}(\cdot)$ represents the edge between robots c_k and c_1. Although, instead of summing up over all edges of the corner hull we can sum up over all robots by rearranging components in l' sum. For each robot c_i we define scalar $l'_i(\alpha_i, \vartheta_i)$ which consist of two components, namely $l'_i(\alpha_i, \vartheta_i) = Pv_i(\alpha_i(1 - \vartheta_i)) + Pv_i(\alpha_i\vartheta_i)$. Then overall speed is represented by

$$l' = \sum_{i=1}^{k} l'_i(\alpha_i, \vartheta_i) = \sum_{i=1}^{k} \cos(\alpha_i\vartheta_i) + \cos(\alpha_i(1 - \vartheta_i)). \tag{6}$$

The function $l'_i(\alpha_i, \vartheta_i)$ is lower bounded by $2\pi^{-2}(\alpha_i - \pi)^2, \forall\alpha_i \in [0, \pi), \forall\vartheta_i \in [0, 1]$. Using this fact we can calculate the lower bound of speed with which length of corner hull decreases, namely

$$l'(t) = \sum_{i=1}^{k} l'_i(\alpha_i, \vartheta_i) \geq \frac{2}{\pi^2} \sum_{i=1}^{k} (\alpha_i - \pi)^2; \tag{7}$$

One of the ways to bound obtained sum is Chebyshev's sum inequality. It is also well known that sum of internal angles of convex polygon is expressed by $\sum_{i=1}^{k} \alpha_i = \pi(k-2)$, where k is the number of polygon's vertices. Thus

$$l'(t) \geq \frac{2}{\pi^2} \left(\frac{1}{k} \left(\pi(k-2) \right) \left(\pi(k-2) - 2\pi \right) \right) + 2k \geq \frac{8}{k}. \tag{8}$$

If all robots belong to the corner hull, then $k = n$ and hull decreases with speed $l'(t) \geq \frac{8}{n}$. □

Lemma 5 states that in 2D the length of the corner hull $l(t)$ decreases continuously with speed $l'(t) \geq \frac{8}{n}$ at any point in time t. This gives us runtime bound $O(l(t_0)n)$, where $l(t_0)$ is length of corner hull around the initial configuration. In the Lemma 6 we are going to show by induction that $l(t_0) \leq 2(n-1)$.

Lemma 6. *The length $l(t_0)$ of the corner hull $CH(R)$ at time $t_0 = 0$ is not greater then $2(n-1)$, where n is a number of robots.*

Proof. We will show that each additional robot can increase the length of the corner hull $l(t_0)$ at time $t_0 = 0$ at most by 2. First assume we have only one robot and $n = 1$, then the length of the corner hull $l(t_0) \leq 2(1-1) = 0$. If we want to increase the length by adding one more robot we need to place it at a maximal possible distance one away from the first one. In this case for $n = 2$ with two robots $l(t_0) \leq 2(2-1) = 2$, since the corner hull consists of two edges of length 1.

Assume that our inequality $l(t_0) \leq 2(k-1)$ holds for arbitrary number of robots k. If we want to increase the length of the corner hull by adding one more robot r to k existing we need to place it outside of existing corner hull $\overline{CH(R)}$ at most distance 1 away from one of the robots c_i that belongs to the hull in order to preserve connectivity.

Let (c_{p_1}, r) and (r, c_{p_2}), where $p_1, p_2 \in k-1$ be new edges of the corner hull around $k+1$ robots. Let $d = |c_{p_1}, r| + |r, c_{p_2}|$ be the length of this edges. When the number of robots was k, length of corner hull between c_{p_1}, c_i was $l(c_{p_1}, c_i) \geq |c_{p_1} c_i|$, and between c_i, c_{p_2} the length was $l(c_i, c_{p_2}) \geq |c_i c_{p_2}|$. Now using triangle inequality we can bound the length of new edges

$$|c_{p_1} r| \leq 1 + |c_{p_1} c_i| \leq 1 + l(c_{p_1} c_i), \tag{9}$$

$$|r c_{p_2}| \leq 1 + |c_i c_{p_2}| \leq 1 + l(c_i c_{p_2}). \tag{10}$$

By summing up inequalities we get the desired result

$$d = |c_{i-p_1} r| + |r c_{i+p_2}| \leq l(c_{i-1} c_i) + l(c_i c_{i+1}) + 2. \tag{11}$$

The difference in length between corner hulls of k and $k+1$ robots is at most 2. □

The runtime bound $O(n^2)$ for the GTC and GTGC algorithms in two-dimensional Euclidean space, follows straight forward from Lemmas 5 and 6.

Theorem 2. *In two-dimensional Euclidean space, GTC algorithm and the GTGC algorithm gather n robots in time $O(n^2)$.*

4   Collisionless Gathering

So far the connectivity property was the only difference that we have seen between GTC and GTGC algorithms. In this section, we present the "collisionless" gathering property that only GTGC algorithm possesses.

At any point in time, if two robots move to the exactly same position, they will share the same position for the rest of the execution of the algorithm since they see the same neighborhood and therefore behave in the same way. We are going to call such event the *collision* of robots. We say that the gathering is *collisionless* if all robots in a group initially had distinct positions and then at the same time moved to the same point. The set of robots M that collide at time t_* is represented as a single robot u for any $t \geq t_*$. We are going to call such robot u the *representative* of M.

Definition 2 (Final collision). *The set of robots M had a final collision at time t if the minimum enclosing circle around the Gabriel neighborhood of the representative u has diameter zero.*

The undesired collision, i.e. *early collision* is just an opposite of final one.

Definition 3 (Early collision). *The set of robots M had an early collision at time t if minimum enclosing circle around the Gabriel neighborhood of the representative u has a diameter greater than zero.*

If there are no early collisions, then the gathering is *collisionless*.

4.1   Collisionless Gathering in One Dimension

Let us consider the group of the robots R that follows GTGC algorithm in one dimension. The initial configuration is connected unit disc graph. Without loss of generality we number the robots in the group with respect to their positions, so that $r_1(t_0) < r_2(t_0) < \ldots < r_n(t_0)$. The positions of all robots are distinct, and the distance between any two consequent robots is less or equal 1.

Robots that execute GTGC algorithm at every point in time calculate their Gabriel graph neighborhood. Let us consider the very first point in time t_0. It is easy to see that the Gabriel graph obtained from unit disc graph is nothing but a path graph with edges $\{r_j(t_0), r_{j+1}(t_0)\}$, where $j = 1, 2, \ldots, n-1$. Each r_i robot might have at most two Gabriel neighbors in one dimension. They are the nearest robots on the right r_{i-1} and on the left r_{i+1} hand side from the robot r_i. All other robots further away will not satisfy Gabriel graph criterion because of those nearest once.

As the next step of GTGC algorithm robots in R calculate their target points with respect to their Gabriel graph neighborhoods. For the robot $r_i \in R$ that has position $r_i(t_0)$, the target point $c_i(t_0)$ is defined in 1D case as follows:

$$c_i(t) = \begin{cases} \frac{r_1(t)+r_2(t)}{2} & \text{if } i = 1; \\ \frac{r_{i-1}(t)+r_{i+1}(t)}{2} & \text{if } i = 2,\ldots,n-1; \\ \frac{r_{n-1}(t)+r_n(t)}{2} & \text{if } i = n. \end{cases} \tag{12}$$

The terminal robots $(i = 1, n)$, i.e. the first one and the last one have only one Gabriel neighbor. Robots move towards their own target points with constant speed 1 and follow the movement of the target points once they have reached them.

Use of GTGC algorithm in one-dimensional space engenders emergent property on the group level. It is the collisionless gathering property that we prove next.

Theorem 3. *In one-dimensional Euclidean space, the gathering with GTGC algorithm is collisionless.*

Proof. Let us consider a group of robots R and subset $M \subset R$. Assume that there was an early collision among members of set M at time $t_1 \in (t_0, t_*)$, where t_* is the time of the final collision.

Let robot u will be representative of M at time t_1. By the definition of the early collision the minimum enclosing circle $C(u)$ around Gabriel neighborhood of u has diameter strictly greater than zero. This means there is at least one more robot v such that $v \notin M$, but $v \in MEC(u)$ at time t_1.

According to Lemma 2 robot v was a Gabriel neighbor of some robot in M. Without loss of generality, we will refer to this robot u as well. The rest of the robots in M might belong to M together with their minimum enclosing sets. But, at least one robot $u \in M$ is such that the robot $v \in MEC(u)$ and $v \notin M$. In other words, the radius of minimum enclosing circle $C(u)$ is strictly positive during the time interval $[t_0, t_1]$. Let ρ be the minimum radius of the circle $C(u)$ during the time interval $[t_0, t_1]$.

There shall be at least two robots in M for an early collision. Let us consider one more robot $w \in M$. According to the assumption of the early collision the limit of the distance between u and w as the time approaches t_1 is equal to zero. There exists point in time $t' \in [t_0, t_1]$, such that $|w(t'), u(t')| < \rho$. Robot w and v shall be on the different sides since otherwise v is not going to be Gabriel neighbor of u during $[t', t_1]$ because of w. Assume without loss of generality v is on the left side from u, then w is on the right. We also may assume without loss of generality that w is a Gabriel neighbor of u. We can do so since if there are any other robots from M between u and w, they all are closer than ρ to u, and the closest one is the Gabriel neighbor of u.

According to the assumption of early collision the limit of the distance between u and w shall be zero as time approaches t_1. However according to GTGC algorithm at any point in time $t' \in [t_0, t_1]$, such that $|w(t'), u(t')| < \rho$

robot u will move away from w with speed 1, since $|v(t'), u(t')| > \rho$. Robot w will not be able to shorten the distance to u due to constant velocity. Thus the distance between u and w will not be less than ρ, which is a contradiction to our assumption of early collision.   □

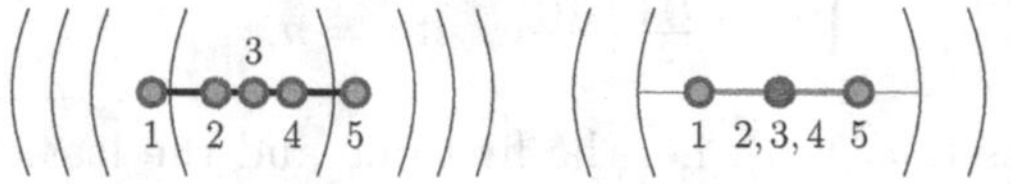

(a) Unit disc graph at time t_0, i.e. initial configuration.   (b) Unit disc graph at t_1.

Fig. 2. An example of the initial configuration that has an early collision, if robots use GTC algorithm. Circles represent robots, red lines represent trajectories of robots, the arcs are parts of according unit discs.

Unlike GTGC, the original GTC algorithm does not have the collisionless gathering property. An example of the configuration that causes an early collision is shown on Fig. 2. At time t_0 robots $1, 5$ belong to the unit graph neighborhood of robot $2, 3$ and 4. Due to this three robots in the middle have the same target point. At time t_1 they have an early collision as it is shown on Fig. 2.

4.2   Almost Collisionless Gathering in Two Dimensions

Does the gathering with GTGC algorithm is also collisionless in two-dimensional Euclidean space? In our experiments, we observe that all randomly generated configurations had no early collisions with GTGC algorithm. An example that highlights the difference behavior of GTC and GTGC is shown in Appendix A.1. This difference we reflect in the following conjecture:

Conjecture 1.   In two-dimensional Euclidean space, the gathering with GTGC algorithm is collisionless for almost every initial configuration.

In other words, we conjecture that with GTGC the set of initial configurations that leads to early collisions has Lebesgue measure zero in configuration space. Besides that, our experiments suggest that configurations that lead to the early collisions with GTGC are highly symmetric. We support our conjecture with measurements obtained from the simulation of the group of robots that executes GTC and GTGC algorithms.

In the simulation, the continuous motion of the robots is replaced by the discrete one with a small discretization step. Two robots have a collision if the distance between them is less than a collision threshold, which is few orders of magnitude less than discretization step. The simulation process is terminated when the maximum distance between any two robots in the group is smaller than the final threshold, which slightly greater than the number of robots times discretization step.

We perform three experiments[1] with different input and measure the number of collisions between the robots during the run. In the first experiment we let robot execute GTC and GTGC on random graphs (e.g. see Fig. 4). The initial configuration, i.e. the random unit disc graph is created by Monte Carlo method. Every robot is placed independently at random on the plane and is removed if it is not connected to the rest of the robots.

The statistical information about the runs of GTC algorithm with the random graphs of different size as an input is presented in Table 1, Appendix A.1. The number of early collisions growth with the size of the input. As well as the deviation, since for every input size among the samples, we have both, configuration with none and many collisions. In the same experiment with the random graph and GTGC algorithm robots had no collisions at all.

In our second experiment, we use an initial configuration for the graph with a specific structure. We aim to show that GTGC actually makes that difference, rather than the input. Here the input is obtained from path graph by placing additional m neighbors very close to every k-th robot in the path, i.e. cluster. We call such graph clustered path graph (see Fig. 5). In this way, we make sure that every such cluster around k-th robot will produce a collision.

The statistical information for the experiments with clustered path graph is presented in the Table 2, Appendix A.1. The low deviation at every input size suggests that GTC with clustered path as input definitely produces early collisions. In the same experiment with clustered path graph as an input GTGC algorithm had no collisions at all.

The only known to us initial configuration that produces early collisions with GTGC algorithms is a cross-shaped graph (see Fig. 6) constructed as follows. The robots are split into two groups H and Q. The robots of one group (e.g. H) are placed along the line h at the equal distance apart from each other. The other group (Q) is placed on the line q perpendicular to h. The line q crosses h at the midpoint between end robots of H on h. The robots of Q are placed on to the distinct positions on q such that the whole graph is connected, and the distance between the end robots of Q no q is less than the distance between end robots of H on h.

In this configuration, all robots in Q will have an early collision at the point where lines q and h cross. If we are going to "shake" once such cross-shaped initial configuration, i.e. every robot moves independently at random in some small δ-ball around its initial position, such that the connectivity is preserved, then the gathering in the simulation with GTGC will also be collisionless.

---

[1] Find a video at: https://youtu.be/jA9foyZegFY.

5   Outlook

In this paper, we have presented and analyzed GTC and GTGC algorithm. We have shown that given algorithms solve the gathering problem in continuous time model for autonomous, oblivious and silent mobile robots with limited visibility in one and two-dimensional Euclidean space. The gathering problem in one dimension is solved in time $\Theta(n)$ and in two dimensions in time $O(n^2)$. Moreover, we have proved that with GTGC algorithm the gathering is collisionless. For two dimensions we carried out experiments which suggest that in the given model the gathering with GTGC is collisionless almost from every initial configuration in two-dimensional Euclidean space.

In the future, we would like to prove almost collisionless gathering in 2D formally. Besides that, we would like to improve obtained upper bound. In order to do it, new analysis techniques would be required. Furthermore, we would like to investigate the impact of different proximity graphs on the solutions of gathering and other similar problems.

A   Appendix

A.1   Examples and experimental results

The Fig. 3 highlights the difference between GTC and GTGC. Initial configuration (Fig. 3(a)) is connected. Figure 3(b) and (c) represent the evolution of the group of the robots that uses GTC algorithm. Figure 3(d) and (e) represent robots that use GTGC algorithm at the same points in time $t_1 < t_2$. Robots $2, 6$ and $3, 5$ that use unit disc graph neighborhood have an early collision at time t_2 and behave as one since that point in time. On the other hand, robots that used GTGC algorithm had no collisions.

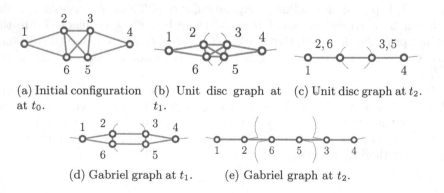

(a) Initial configuration at t_0.

(b) Unit disc graph at t_1.

(c) Unit disc graph at t_2.

(d) Gabriel graph at t_1.

(e) Gabriel graph at t_2.

Fig. 3. Visibility graphs during the gathering of the robots that use original GTC and modified GTGC. Blue circles represent robots; red lines represent trajectories of robots.

The Figs. 4, 5, 6 illustrate instances of the graphs used in the experiments.

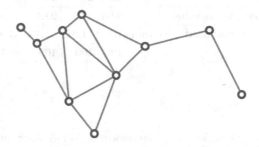

Fig. 4. An instance of the random unit Gabriel graph

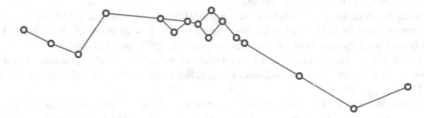

Fig. 5. An instance of the random clustered path graph with Gabriel edges and a single cluster in the middle.

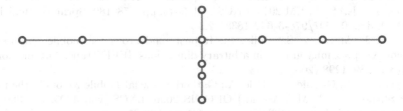

Fig. 6. An instance of the cross shaped graph that leads to the early collisions with GTGC

Table 1. GTC algorithm with the random unit disc graph as input

Number of robots		10	20	30	40	50	60
Number of collisions	Mean value	0.18	1.87	3.52	8.45	15.41	19.7
	Standard deviation	0.67	3.19	4.55	8.38	11.57	14.29
Sample size		100	100	100	100	100	100

Table 2. GTC algorithm with a clustered path graph as an input

Number of robots		16	26	36	46	56	66
Number of collisions	Mean value	9.05	17.7	24.1	33.85	43.6	54.75
	Standard deviation	1.1	1.69	1.45	1.81	1.24	1.29
Sample size		100	100	100	100	100	100

References

1. Agathangelou, C., Georgiou, C., Mavronicolas, M.: A distributed algorithm for gathering many fat mobile robots in the plane. In: Proceedings of the 2013 ACM Symposium on Principles of Distributed Computing, PODC 2013, pp. 250–259. ACM, New York, NY, USA (2013)
2. Ando, H., Suzuki, I., Yamashita, M.: Formation and agreement problems for synchronous mobile robots with limited visibility. In: Proceedings of the 1995 IEEE International Symposium on Intelligent Control, 1995, pp. 453–460 (1995)
3. Chrystal, G.: On the problem to construct the minimum circle enclosing n given-points in a plane. In: Proceedings of the Edinburgh Mathematical Society, Third Meeting, pp. 30–35 (1885)
4. Cohen, R., Peleg, D.: Convergence properties of the gravitational algorithm in asynchronous robot systems. In: Albers, S., Radzik, T. (eds.) ESA 2004. LNCS, vol. 3221, pp. 228–239. Springer, Heidelberg (2004). doi:10.1007/978-3-540-30140-0_22
5. Cord-Landwehr, A., et al.: Collisionless gathering of robots with an extent. In: Černá, I., Gyimóthy, T., Hromkovič, J., Jefferey, K., Královič, R., Vukolić, M., Wolf, S. (eds.) SOFSEM 2011. LNCS, vol. 6543, pp. 178–189. Springer, Heidelberg (2011). doi:10.1007/978-3-642-18381-2_15
6. Cortes, J., Martinez, S., Bullo, F.: Robust rendezvous for mobile autonomous agents via proximity graphs in arbitrary dimensions. IEEE Trans. Autom. Control **51**(8), 1289–1298 (2006)
7. Czyzowicz, J., Gsieniec, L., Pelc, A.: Gathering few fat mobile robots in the plane. In: Shvartsman, M.M.A.A. (ed.) OPODIS 2006. LNCS, vol. 4305, pp. 350–364. Springer, Heidelberg (2006). doi:10.1007/11945529_25
8. Degener, B., Kempkes, B., Langner, T., auf der Heide, F.M., Pietrzyk, P., Wattenhofer, R.: A tight runtime bound for synchronous gathering of autonomous robots with limited visibility. In: Proceedings of the 23rd ACM Symposium on Parallelism in Algorithms and Architectures, SPAA 2011, pp. 139–148. ACM, New York, NY, USA (2011)
9. Ruben Gabriel, K., Sokal, R.: A new statistical approach to geographic variation analysis. Syst. Biol. **18**(3), 259–278 (1969)
10. Gordon, N., Wagner, I.A., Bruckstein, A.M.: Gathering multiple robotic a(ge)nts with limited sensing capabilities. In: Dorigo, M., Birattari, M., Blum, C., Gambardella, L.M., Mondada, F., Stützle, T. (eds.) ANTS 2004. LNCS, vol. 3172, pp. 142–153. Springer, Heidelberg (2004). doi:10.1007/978-3-540-28646-2_13
11. Karp, B., Kung, H.T.: Gpsr: Greedy perimeter stateless routing for wireless networks. In: Proceedings of the 6th Annual International Conference on Mobile Computing and Networking, MobiCom 2000, pp. 243–254. ACM, New York, NY, USA (2000)

12. Kempkes, B., Kling, P., auf der Heide, F.M.: Optimal and competitive runtime bounds for continuous, local gathering of mobile robots. In: Proceedinbgs of the 24th ACM Symposium on Parallelism in Algorithms and Architectures, SPAA 2012, pp. 18–26. ACM, New York, NY, USA (2012)
13. Megiddo, N.: Linear-time algorithms for linear programming in $\mathbb{R}^3$ and related problems. SIAM J. Comput. **12**(4), 759–776 (1983)
14. Pagli, L., Prencipe, G., Viglietta, G.: Getting close without touching. In: Even, G., Halldórsson, M.M. (eds.) SIROCCO 2012. LNCS, vol. 7355, pp. 315–326. Springer, Heidelberg (2012). doi:10.1007/978-3-642-31104-8_27

Search-and-Fetch with One Robot on a Disk
(Track: Wireless and Geometry)

Konstantinos Georgiou[1], George Karakostas[2(✉)], and Evangelos Kranakis[3]

[1] Department of Mathematics, Ryerson University, Toronto, ON, Canada
[2] Department of Computing and Software,
McMaster University, Hamilton, ON, Canada
karakos@mcmaster.ca
[3] School of Computer Science, Carleton University, Ottawa, ON, Canada

Abstract. A robot is located at a point in the plane. A treasure and an exit, both stationary, are located at unknown (to the robot) positions both at distance one from the robot. Starting from its initial position, the robot aims to fetch the treasure to the exit. At any time the robot can move anywhere on the disk with constant speed. The robot detects an interesting point (treasure or exit) only if it passes over the exact location of that point. Given that an adversary controls the locations of both the treasure and the exit on the perimeter, we are interested in designing algorithms that minimize the treasure-evacuation time, i.e. the time it takes for the treasure to be discovered and brought to the exit by the robot.

In this paper we differentiate how the robot's knowledge of the distance between the two interesting points affects the overall evacuation time. We demonstrate sthe difference between knowing the exact value of that distance versus knowing only a lower bound and provide search algorithms for both cases. In the former case we give an algorithm which is off from the optimal algorithm (that does not know the locations of the treasure and the exit) by no more than $\frac{4\sqrt{2}+3\pi+2}{6\sqrt{2}+2\pi+2} \leq 1.019$ multiplicatively, or $\frac{\pi}{2} - \sqrt{2} \leq 0.157$ additively. In the latter case we provide an algorithm which is shown to be optimal.

Keywords: Disk · Exit · Robot · Search and Fetch · Treasure

1 Introduction

Search is concerned with finding an object under various conditions within a search space. In the context of computational problems this usually becomes more challenging especially when the environment is unknown to the searcher (see [1,3,24]) and efficient algorithms with respect to search time are sought. For example, in robotics exploration may be taking place within a given geometric

Research supported in part by NSERC Discovery grants.

M. Chrobak et al. (Eds.): ALGOSENSORS 2016, LNCS 10050, pp. 80–94, 2017.
DOI: 10.1007/978-3-319-53058-1_6

domain by a group of autonomous but communicating robots and the ultimate goal is to design an algorithm so as to accomplish the requirements of the search (usually locating a target of unknown a priori position) while at the same time obeying the computational and geographical constraints. Further, the task must be accomplished in the minimum possible amount of time [8].

There is extensive research and several models have been proposed and investigated in the mathematical and theoretical computer science literature with particular emphasis on probabilistic search [24], game theoretic applications [3], cops and robbers [9], classical pursuit and evasion [23], search problems as related to group testing [1], searching a graph [22]. A survey of related search and pursuit evasion problems can be found in [11], whereby pursuers want to capture evaders trying to avoid capture. Examples include *Cops and Robbers* (where the cops try to capture the robbers by moving along the vertices of a graph), *Lion and Man* (a geometric version of cops and robbers where a lion is to capture a man in either continuous or discrete time), etc. Searching for a stationary point target has some similarities with the lost at sea problem, [17,18], the cow-path problem [6,7], and with the plane searching problem [5].

In this paper, we study a new problem which involves a robot *searching* for a treasure and *fetching* it to an exit. Both treasure and exit are at distance 1 from the starting position of the robot (i.e., located on the perimeter of a unit radius circle) at locations unknown to the robot. The robot can move with maximum speed 1, starts at the centre of a circle and continues by moving to the perimeter. The adversary has control over the locations of both the exit and the treasure. Goal is to provide algorithms that minimize the "search time" for the robot to find the treasure and bring it to the exit. Surprisingly, finding an optimal algorithm turns out to be a rather difficult problem even when the robot has some knowledge on the arc-distance between exit and treasure.

There are several problems in the scientific literature relating to evacuation, although of very different nature than our problem. For grid polygons, evacuation has been studied in [16] from the perspective of constructing centralized evacuation plans, resulting in the fastest possible evacuation time from the rectilinear environment. Our problem has similarities to the well-known evacuation problem on an infinite line (see [4] and the more recent [10]) in that the search is for an unknown target; an important difference is that, in the basic optimal zigzag algorithm presented in [4], the search is on an infinite line which limits the possibilities for the adversary. Additional research and variants on this problem can be found in [15] (on searching with turn costs), [21] (randomized algorithm for the cow-path problem), and [20] (hybrid algorithms).

Our model is relevant to the recent works [12–14] investigating algorithms in the wireless and non-wireless (or face-to-face) communication models for the evacuation of a team of robots. Note that in this case, the "search domain" is the same and the evacuation problem without a treasure for a single robot is trivial. Thus, in addition to searching for the two stationary objects (namely treasure and exit) at unknown locations in the perimeter of a cycle we are also interested in fetching the treasure to the exit. As such, our search-and-fetch type problem is of much different nature than the series of evacuation-type problems above, and in fact solutions to our problem require a novel approach.

Our optimization problem models real-life situations that may arise in surveillance, emergency response, and search-and-rescue operations, e.g. by aerial drones or other unmanned vehicles. Indeed, consider a rescue-robot that receives a distress signal indicating that a victim is at an unknown location but at known distance from a safe shelter. What is the optimal trajectory of the robot that can locate the victim and fetch it to the shelter? Similar problems are well studied in the robotics community since the 90's, e.g. see [19]. A search-and-fetch problem similar to ours was introduced by Alpern in [2], where the underlying domain was discrete and the approach/analysis resembled that of standard search-type problems [3]. In contrast, the focus of the current work is to demonstrate how some knowledge of the input may affect optimality in designing online solutions when there is only one rescue-robot available.

2   Preliminaries, Notation, and Results of the Paper

We begin with presenting the precise definitions of the treasure evacuation problem and some basic notation, concepts and necessary definitions.

A treasure and an exit are located at unknown positions on the perimeter of a unit-disk and at arc distance α (in what follows all distances will be arc-distances, unless specified otherwise). A robot starts from the center of the disk, and can move anywhere on the disk at constant speed 1. The robot detects the treasure or the exit only if its trajectory passes over that point on the disk. Once detected, the treasure can be carried by the robot at the same speed. Our goal is to design algorithms that minimize the evacuation time, i.e. the time it takes from the moment the robot starts moving, till the treasure is detected and brought to the exit by the robot. Sometimes we refer to the task of bringing the treasure to the exit as *treasure-evacuation*. We also use the abbreviations T, E for the treasure and the exit, respectively. For convenience, in the sequel we will refer to the locations of the exit and the treasure as *interesting* points. For an interesting point I on the perimeter of the disk, we also write $I = E$ ($I = T$) to indicate that the exit (treasure) lies at point I.

We focus on the following two variants of treasure-evacuation reflecting the knowledge of the robot with respect to its environment.

Definition 1. *In **1-TE$_=$**, one robot attempts treasure-evacuation knowing that the distance between T, E is exactly α. In **1-TE$_\geq$**, one robot attempts treasure-evacuation knowing that the distance between T, E is at least α.*

2.1   Final Steps of Treasure-Evacuation

The exploration part of any evacuation algorithm concludes with the discovery of an interesting point (i.e., treasure or exit). This leads us to define the concepts of *double* and *triple* move that will prove useful in the subsequent analysis for the case when the robot knows that the exit and the treasure are at distance exactly

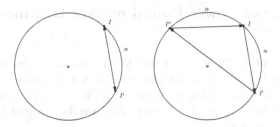

Fig. 1. The double move (left) and triple move (right) of a robot occur when the first interesting point to be encountered by the robot is an exit.

α. If a robot encounters an interesting point at I then the other interesting point is either at point I' or I'' (Fig. 1, right), both at arc distance α from I.

If the robot has already explored one of these points (say I'') without finding an interesting point there, then the situation reduces to the left of Fig. 1. If this is the robot that will eventually evacuate the treasure, then the worst case is to find the exit in I and the treasure in I'; in this case, the robot will perform a *double move*, going to I', pick up the treasure, and return to I, spending time equal to twice the chord distance between I and I', i.e. $4\sin(\alpha/2)$. On the other hand, if both I', I'' are still unexplored, the worst case for the algorithm would be to perform a *triple move*, i.e., find the exit at I, then visit I' without finding anything there, then visit I'' and pick up the treasure, and return to I, always moving along the shortest underlying chords, inducing time $4\sin(\alpha/2) + 2\sin(\alpha)$. Once an interesting point is found, we will refer to the above process as the evacuation step. Clearly, this will involve either a double or a triple move, depending on whether sufficiently many points on the disk have already been explored.

2.2 Outline and Results of the Paper

We study treasure evacuation for a single robot. We contrast how knowledge affects the evacuation time by considering two models: in the first the robot knows only a lower bound on the actual arc-distance, while in the second the robot knows the distance between treasure and exit. When a lower bound α on the actual arc-distance l is known to the robot (see Sect. 3) then we give an optimal treasure evacuation algorithm which takes time $1 + 2\pi - \alpha + 2\sin(\alpha/2) + 2\sin(l/2)$. When the robot knows that α is not just a lower bound, but the *actual* arc-distance between treasure and exit (i.e., $l = \alpha$) (see Sect. 4), then we propose the *arc-partition* algorithm which makes the robot alternate between exploring and hopping arcs and use continuous optimization techniques in order to show that it is nearly optimal. More specifically, our upper and lower bounds are off multiplicatively by no more than $\frac{4\sqrt{2}+3\pi+2}{6\sqrt{2}+2\pi+2} \le 1.019$, when $\alpha = \pi/2$, while for $\alpha \to 0$ or $\alpha \to \pi$ our algorithm is nearly optimal. In Sect. 5 we conclude and suggest several extensions and open problems.

3 Knowledge of Lower Bound on Arc Distance: Problem 1-TE$_\geq$

In this section we prove tight upper and lower bounds for treasure evacuation when only a lower bound, say α, on the arc distance between exit and treasure is known to the robot. Note that for problem 1-TE$_\geq$ considered here the special case $\alpha = 0$ means that the robot does not know anything about the arc distance between treasure and exit.

Our "Arc Avoidance" Evacuation Algorithm is given below. As implied by its name, the algorithm merely avoids exploration of an arc of length α following the encounter of the first interesting point.

Algorithm 1. Arc Avoidance Evacuation Algorithm

Step 1. Starting at the center of the circle, move to the perimeter and start exploring Clock-Wise (CW) along the perimeter until the first interesting point A is found.

Step 2. Move CW along chord AB of length $2\sin(\alpha/2)$ (see Fig. 2).

Step 3. Continue exploring from point B on the perimeter until the second interesting point is found.

Step 4. Evacuate the treasure using either a double or a triple move.

Theorem 1. *When the robot is given a lower bound α on the actual arc distance l between exit and treasure, the evacuation time of Algorithm 1 is at most $1 + 2\pi - \alpha + 2\sin(\alpha/2) + 2\sin(l/2)$ in the worst-case. Further, Algorithm 1 is worst-case optimal, i.e. no algorithm can attain a better time in the worst case for a robot to evacuate the treasure.*

Proof. (Theorem 1) We prove the upper and lower bounds separately.

3.1 Upper Bound

Let x be the time it takes until the robot encounters at A the first interesting point on the perimeter of the disk. Consider the two cases depicted in Fig. 2.

Case 1: $x > \alpha$. This is depicted in Fig. 2 (left). The robot makes the move $A \to B$ along the chord. Then it explores again the perimeter. Let y be the time until it finds the second interesting point at C. It is clear that the worst total cost (i.e., $A = E, C = T$) is

$$1 + x + 2\sin(\alpha/2) + y + 2\sin(l/2), \tag{1}$$

where $2\sin(l/2)$ is the length of the chord CA connecting the treasure to the exit. Further, observe that $y \leq 2\pi - \alpha - x$, which implies that $x + y \leq 2\pi - \alpha$. Also, for this case we have that $x \geq \alpha$ and, since $y \leq 2\pi - \alpha - x$, the bound follows.

Case 2: $x \leq \alpha$. This is depicted in Fig. 2 (right). Let B, B' be the two points at arc distance α from A, and y be the time it takes for the robot to find the second

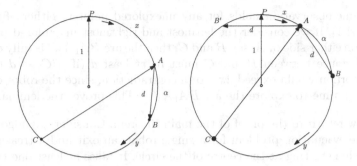

Fig. 2. The two cases of the evacuation Algorithm 1.

interesting point at C. As before, the evacuation cost is given by Formula (1). Observe that in this case the robot does not need to traverse the arc $\overgroup{B'AB}$ (by assumption the other interesting point must be at arc distance at least α) and therefore $y \leq 2\pi - 2\alpha$. Also since $x \leq \alpha$ we also have $x + y \leq 2\pi - \alpha$, and the bound follows.

3.2 Lower Bound

First we prove the following lemma which will be used in the proof of the lower bound below.

Lemma 1. *If a robot during its exploration of the perimeter has covered length less than $2\pi - \alpha$ then there is a chord of length exactly $2\sin(\alpha/2)$ none of whose endpoints has been explored by the robot.*

Proof. (Lemma 1) Let us define $d := 2\sin(\alpha/2)$. Assume on the contrary no such chord exists. It follows that for any unexplored point A if we draw two chords AA_1 and AA_2 each of length d then each point in the arc A_1A_2 must have been explored by the robot (see Fig. 3).

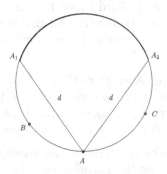

Fig. 3. Proving that there is a chord of length at least d with two unexplored endpoints.

The same observation holds for any unexplored point in either of the arcs AA_1 and AA_2. If we consider the leftmost and rightmost unexplored points on the circle on either side of A, say B and C, then the arc BA_1A_2C is fully explored and the distance between B and C must be at least d (if $|BC| < d$ then the explored portion would exceed $2\pi - \alpha$, a contradiction, since the robot does not have enough time to explore the arc BA_1A_2C). This proves the lemma.   □

We now return to the proof of the main theorem. Consider any algorithm $\mathcal{A}$ solving the evacuation problem for a single robot, an exit and a treasure. Run the algorithm starting at the centre of the circle. It takes at least one time unit for the robot to reach the perimeter. The robot starts exploring the perimeter after this time. Run algorithm $\mathcal{A}$ for additional $2\pi - \alpha - \epsilon$ time. During this additional time, the robot can explore a total length of at most $2\pi - \alpha - \epsilon$ of the perimeter.

Take a pair of points delimiting a chord with the properties specified in Lemma 1. Such a chord has length exactly $d = 2\sin(\alpha/2)$ and has both endpoints unexplored. Clearly, the robot has not yet found neither the exit nor the treasure. Following the algorithm $\mathcal{A}$ the robot will visit one of its endpoints first; the adversary can place the exit at this first endpoint and the treasure in the other. Therefore it will take additional time at least $2\sin(l/2)$ to evacuate the treasure, and the total evacuation time will be at least $1 + 2\pi - \alpha - \epsilon + 2\sin(\alpha/2) + 2\sin(l/2)$, for any $\epsilon > 0$. This completes the proof of Theorem 1.   □

4   Exact Knowledge of Arc-Distance: Problem 1-TE$_=$

In this section, we separate our analysis into an upper bound accompanied by a search algorithm and a lower bound using continuous optimization techniques.

4.1   Upper Bound

When a robot knows the exact arc-distance of the interesting points, there is a naive evacuation protocol, which we call the *Sweeping Algorithm*: go to an arbitrary point on the perimeter of the circle, and start traversing clockwise till an interesting point is found. Then either perform a double or a triple move to finalize the evacuation. It is not hard to see that in the worst case, this naive algorithm has cost $1 + 2\pi - \alpha + 4\sin(\alpha/2)$. Our goal in this section is to improve upon this naive approach. To explain the main evacuation algorithm we first provide a definition.

Definition 2 (Alternating Arcs Partition). *An alternating arcs partition is a partition $\mathcal{P}$ of the perimeter of the cycle into $2n + 1$, $n \geq 0$ consecutive and pairwise non-intersecting arcs $a_1, b_1, a_2, b_2, \ldots, a_n, b_n, a_{n+1}$, such that $b_i \leq \alpha$, $\forall i$. If in addition we have that $\alpha \leq a_i$, we call the partition α-greedy.*

Our Algorithm 2 first picks an alternating arc partitioning $\mathcal{P}$, before it follows through with its steps. In case the partition contains only a_1 (i.e., $n = 0$), the algorithm performs no "jumps" along chords until the first interesting point is encountered, i.e. it is exactly the Sweeping Algorithm.

Algorithm 2. Alternating Arcs, on input partition $\mathcal{P}$

Step 1. Walk to the perimeter, say point P;

Step 2. Keep moving in CW direction *arc-sweeping* each a_i, and *cord-jumping* along the chord of each b_i, until the first interesting point is found.

Step 3. Evacuate the treasure using either a double or a triple move.

Algorithm 2 succeeds in evacuating the treasure when run on input some α-greedy partition. Correctness is an immediate corollary of the definition of the partition (that depends on the value of α). Note that the *chord-jumps* correspond to arcs of size at most α, hence two interesting points cannot fit in any arc b_i. Also, since the *sweeping-arcs* a_i have length at least α, two interesting points cannot be in two (consecutive) arcs b_i, b_{i+1}. It follows that, eventually, a robot following Algorithm 2 will locate an interesting point and will evacuate with either a double or a triple move.

It remains to define the partition $\mathcal{P}$ used by Algorithm 2. The intuition behind our partition is based on the following greedy rule: perform as many alternating arc-chord moves as possible so that the worst configuration of interesting points which concludes evacuation with a double-move has total time no more than the worst configuration of interesting points which concludes evacuation with a triple-move.

Definition 3 (Greedy Partition $\mathcal{P}_\alpha$). *For every $\alpha > 0$, we set*

$$\kappa_\alpha := \left\lfloor \frac{2\pi - 3\alpha - 2\sin(\alpha)}{2\alpha} \right\rfloor, \quad slack_\alpha := 2\pi - (2\kappa_\alpha + 3)\alpha - 2\sin(\alpha)$$

and

$$\gamma_\alpha := \begin{cases} \alpha & , if\ slack_\alpha > \alpha \\ slack_\alpha & , otherwise \end{cases}, \quad \chi_\alpha := \begin{cases} slack_\alpha - \alpha & , if\ slack_\alpha > \alpha \\ 0 & , otherwise \end{cases}$$

Then, we define

$$\mathcal{P}_\alpha := \begin{cases} a_1, b_1, \ldots, a_{\kappa_\alpha}, b_{\kappa_\alpha}, a_{\kappa_\alpha+1}, b_{\kappa_\alpha+1}, a_{\kappa_\alpha+2} & , if\ \kappa_\alpha \geq 0 \\ 2\pi & , otherwise \end{cases}$$

where $a_i = b_i = \alpha$, for $i = 1 \ldots \kappa_\alpha$, $a_{\kappa_\alpha+1} = \alpha$, $b_{\kappa_\alpha+1} = \gamma_\alpha$, and $a_{\kappa_\alpha+2} = 2\alpha + 2\sin(\alpha) + \chi_\alpha$.

It is clear that if $\kappa_\alpha < 0$, then the above partition gives rise to the sweeping algorithm, which happens after the root $\alpha_0 \approx 1.43396$ of the equation $2\pi = 3\alpha + 2\sin(\alpha)$. It is also clear that $\mathcal{P}_\alpha$ is an α-greedy alternating arcs partition.

Next we show that the worst configuration for Algorithm 2 that uses partition $\mathcal{P}_\alpha$ is indeed a double-move.

Lemma 2. *The worst case configuration for Algorithm 2 that uses partition $\mathcal{P}_\alpha$ makes the robot perform a double-move. An upper bound for the cost can be* *computed by placing the treasure arbitrarily close to the starting point P (from the opposite direction than the one that the robot starts sweeping) and the exit at CCW-arc-distance α from it.*

Proof. The statement is true for the sweeping algorithm, since the worst-case double-move is no less costly than the worst triple-move. So we may focus on the case slack$_\alpha > 0$, in which we do have at least one chord-move.

We observe that after the last chord-move, the robot is at some point A at distance $2\alpha + 2\sin(\alpha) + \chi_\alpha$ away from returning to the original point P. If it is forced to do any double or triple move after point A, that would make the total cost at least as costly as in any other double or triple move (respectively) configuration before that.

Next recall that the cost of the double-move alone is $4\sin(\alpha/2)$, while that of a triple-move (alone) is $4\sin(\alpha/2) + 2\sin(\alpha)$. The worst-case scenario ending with a double-move would be to have the treasure close to P and the exit at distance α from P in a Counter-CW (CCW) direction; by construction, the robot will have to sweep an additional arc of length $\alpha + 2\sin(\alpha) + \chi_\alpha - \epsilon$ before encountering an interesting point, i.e. before the double-move starts. In contrast, the worst-case triple-move scenario would be possible only if the robot discovered an interesting point in distance $\alpha - \epsilon$ after point A. The fact that χ_α is always non negative shows the claim. This proves Lemma 2.   □

Now we can give an upper bound to the performance of our algorithm.

Theorem 2. *For every $\alpha > 0$, the performance of Algorithm 2 that uses partition $\mathcal{P}_\alpha$ is at most*

$$\begin{cases} 2\pi - (\kappa_\alpha + 2)\alpha + 2(\kappa_\alpha + 3)\sin(\alpha/2), & \text{if } \pi > (\kappa_\alpha + 2)\alpha + \sin\alpha \\ (\kappa_\alpha + 2)\alpha + 2(\kappa_\alpha + 2)\sin(\alpha/2) + 2\sin\left(\frac{2\kappa_\alpha + 3}{2}\alpha + \sin\alpha\right) + 2\sin\alpha, & o.w. \end{cases}$$

where $\kappa_\alpha = \left\lfloor \frac{2\pi - 3\alpha - 2\sin(\alpha)}{2\alpha} \right\rfloor$.

Proof. By Lemma 2, the worst configuration will place the treasure close to P and the exit at distance α from P in a CCW direction. Hence, using the α-greedy partition $\mathcal{P}_\alpha$, the cost of Algorithm 2 is $1 + \sum_{i=1}^{\kappa_\alpha+1} a_i + 2\sum_{i=1}^{\kappa_\alpha+1} \sin(b_i/2) + (a_{\kappa_\alpha+2} - \alpha) + 4\sin(\alpha/2)$. Using Definition 3, we can expand the above formula with respect to whether $\kappa_\alpha \geq 0$ and slack$_\alpha > \alpha$ or not. Observing that $\kappa_\alpha \geq -1$ for all α and by simplifying the resulting formula yields the promised expression. Note that when $\kappa_\alpha = -1$, the above formula indeed induces the cost of the sweeping algorithm.   □

4.2   Lower Bound

Next we prove our main lower bound for the problem 1-TE$_=$. In general, arguments for lower bounds are based on an adversary detecting the input configuration that maximizes the worst-case time of an optimal algorithm (for that configuration). We are not able to analyze such an all-powerful adversary (which would give us the best (i.e., highest) lower bound). Instead, we analyze an adversary that is restricted as follows to place the exit and the treasure only when the algorithm has left only one completely unexplored pair of points A, B at distance α, and only at the points A and B, therefore obtaining a weaker lower bound.

Theorem 3. *Any algorithm that solves problem 1-TE$_=$ must run for time at least*

$$1 + \pi + \min \left\{ \begin{array}{l} 4\sin\frac{\alpha}{2} + 2\left(\left\lceil\frac{\pi}{\alpha}\right\rceil - 1\right)\sin\frac{\pi-\alpha}{2\left(\left\lceil\frac{\pi}{\alpha}\right\rceil-1\right)}, \\ \pi - \alpha\left\lfloor\frac{\pi}{\alpha}\right\rfloor + 2\left(\left\lfloor\frac{\pi}{\alpha}\right\rfloor + 1\right)\sin\frac{\alpha}{2} \end{array} \right\}$$

Proof. We restrict the adversary to place the exit and the treasure only when the algorithm has left only one completely unexplored pair of points A, B at distance α, and only at the points A and B. Obviously, with such a restricted adversary and because of the triangle inequality, the optimal algorithm will follow an alternating arcs partition in a single direction (say CW), starting at A, ending its exploration phase at B, and leaving the arc AB of length α completely unexplored; at this point, the adversary will have placed the treasure at A and the exit at B, forcing a double-move.

Therefore, we have that $1 + \sum_{i=1}^{l} a_i + \sum_{i=1}^{l-1} 2\sin(b_i/2) + 4\sin(\alpha/2)$ is a lower bound on the evacuation time of an algorithm that chooses an alternating arcs partition, leaving a single unexplored segment b_l, and then performs a double move. Since $1 + 4\sin(\alpha/2)$ doesn't depend on the lengths of a_i, b_i, we get a lower bound for the optimal algorithm by minimizing $\sum_{i=1}^{l} a_i + \sum_{i=1}^{l-1} 2\sin(b_i/2)$, given the constraints we have imposed so far. This leads to the following family of optimization problems (one for each integer $l \geq 1$):

$$\text{minimize} \quad \sum_{i=1}^{l} a_i + \sum_{i=1}^{l-1} 2\sin(b_i/2) \qquad \text{(MP)}$$

$$\text{subject to} \quad \sum_{i=1}^{l} a_i + \sum_{i=1}^{l} b_i = 2\pi \qquad (2)$$

$$b_i \leq \alpha \quad \text{for } i = 1, 2, \ldots, l \qquad (3)$$

$$a_i, b_i \geq 0 \quad \text{for } i = 1, 2, \ldots, l \qquad (4)$$

which is equivalent to

$$\text{minimize} \quad -\sum_{i=1}^{l} b_i + 2\sum_{i=1}^{l-1} \sin(b_i/2) \qquad \text{(MP')}$$

$$\text{subject to } b_i \leq \alpha \quad \text{for } i = 1, 2, \ldots, l \qquad (5)$$

$$b_i \geq 0 \quad \text{for } i = 1, 2, \ldots, l \qquad (6)$$

The overall lower bound we will calculate is the minimum amongst the candidate lower bounds calculated for each value of $l \geq 1$. Note that for $l = 1$, the bound corresponds to the case of a sweeping algorithm with $a_1 = 2\pi - b_1$. Given our adversary assumption, we have $b_1 = \alpha$. In this case the candidate lower bound for the running time is

$$LA = 1 + 4\sin(\alpha/2) + 2\pi - \alpha. \tag{7}$$

We estimate now the candidate lower bounds $LB(l)$ for $l \geq 2$ using (MP'). For brevity reasons, we set $x := \sum_{i=1}^{l} a_i$. Constraint (2) implies that $x \leq 2\pi - b_l$. Also, we have $\pi \leq x$, because of the following claim:

Claim. $\sum_{i=1}^{l} a_i \geq \sum_{i=1}^{l} b_i$.

Proof. Moving clockwise, and starting from b_1, we can map each point of $b_1, b_2, \ldots$ to a unique explored point of some a_i at distance α clockwise, since there is only one pair of unexplored points in distance α when the first interesting point is found (at one of these two points). Therefore, the total length of the b_i's is mapped one-to-one to the a_i's, and the claim follows.   □

For the special case $x = 2\pi - b_l$, we have that $b_i = 0, i = 1, \ldots, l-1$ (from (2)), and the bound becomes the bound in (7) (since all the a_i's form a single contiguous explored segment and then $b_l = \alpha$, according to our adversary assumption). Therefore we can assume that

$$\pi \leq x < 2\pi - b_l. \tag{8}$$

We note that
$$\frac{2\pi - b_l - x}{l-1} \leq \alpha \Leftrightarrow l \geq 1 + \frac{2\pi - x - b_l}{\alpha}, \tag{9}$$

because, otherwise, $2\pi > (l-1)\alpha + b_l + x \geq \sum_{i=1}^{l} a_i + \sum_{i=1}^{l} b_i = 2\pi$, a contradiction. We distinguish two cases:

Case 1: $\frac{2\pi - x - b_l}{l-1} < \alpha$

Suppose that an optimal solution of (MP') has a pair of optimal values $b_i < \frac{2\pi - b_l - x}{l-1}$ and $b_j > \frac{2\pi - b_l - x}{l-1}$ (note that if there is such a b_i, there must be such a b_j, and vice versa, due to (2)). Then there is $\varepsilon > 0$ such that $b_i + \varepsilon < \frac{2\pi - b_l - x}{l-1}$ and $b_j - \varepsilon > \frac{2\pi - b_l - x}{l-1}$. Notice that

$$\sin\left(\frac{b_i + \varepsilon}{2}\right) + \sin\left(\frac{b_j - \varepsilon}{2}\right) < \sin\left(\frac{b_i}{2}\right) + \sin\left(\frac{b_j}{2}\right),$$

since the LHS is a monotonically decreasing function of ε, a contradiction since by setting b_i, b_j to $b_i + \varepsilon, b_j - \varepsilon$ we get a better feasible solution of (MP') than the optimal. Hence

$$b_i = \frac{2\pi - b_l - x}{l-1}, \quad i = 1, \ldots, l-1 \tag{10}$$

and (MP') is equivalent of solving the following minimization problem

$$\min_{\pi \leq x < 2\pi - b_l} x + 2(l-1)\sin\frac{2\pi - b_l - x}{2(l-1)}.$$

If we set $y := \frac{2\pi - b_l - x}{2(l-1)}$, the latter is equivalent to solving

$$\min_{0 < y \leq \frac{\pi - b_l}{2(l-1)}} -b_l - 2(l-1)y + 2(l-1)\sin y.$$

For any value of b_l, the optimal solution has to minimize

$$\min_{0 < y \leq \frac{\pi - b_l}{2(l-1)}} \sin y - y$$

and, therefore, $y = \frac{\pi - b_l}{2(l-1)}$ in the optimal solution. This also implies that $x = \pi$. Hence, the objective function becomes a function of b_l:

$$\min_{0 \leq b_l \leq \alpha} 2(l-1)\sin\frac{\pi - b_l}{2(l-1)}.$$

Since $\frac{\pi - b_l}{2(l-1)} \leq \frac{\pi}{2}$, the minimum is achieved for $b_l = \alpha$. Therefore $b_i = \frac{\pi - \alpha}{l-1}$, $i = 1, \ldots, l-1$ (note that $\frac{\pi - \alpha}{l-1} \leq \alpha$ because of (9)). In this case, the lower bound candidate is

$$LB(l) = 1 + 4\sin\frac{\alpha}{2} + \pi + 2(l-1)\sin\frac{\pi - \alpha}{2(l-1)}. \tag{11}$$

(11) is increasing in l, and, therefore, the smallest $LB(l)$ is (because of (9))

$$LB\left(\left\lceil\frac{\pi}{\alpha}\right\rceil\right) = 1 + 4\sin\frac{\alpha}{2} + \pi + 2\left(\left\lceil\frac{\pi}{\alpha}\right\rceil - 1\right)\sin\frac{\pi - \alpha}{2(\lceil\frac{\pi}{\alpha}\rceil - 1)} \tag{12}$$

if $\frac{\pi}{\alpha} \notin \mathbb{N}$, and

$$LB\left(\frac{\pi}{\alpha} + 1\right) = 1 + 4\sin\frac{\alpha}{2} + \pi + \frac{2\pi}{\alpha}\sin\frac{\alpha(\pi - \alpha)}{2\pi} \tag{13}$$

if $\frac{\pi}{\alpha} \in \mathbb{N}$.

Case 2: $\frac{2\pi - x - b_l}{l-1} = \alpha \Leftrightarrow x = 2\pi - b_l - \alpha(l-1)$

In this case, (5) implies that $b_i = \alpha, i = 1, \ldots, l-1$. Also, (8) implies that $b_l \leq \pi - \alpha(l-1)$. Since $b_l \geq 0$, this implies that $l \leq 1 + \frac{\pi}{\alpha}$. The candidate lower bounds (parameterized by l) are

$$LC(l) = 1 + 4\sin\frac{\alpha}{2} + 2\pi - b_l - \alpha(l-1) + 2(l-1)\sin\frac{\alpha}{2}.$$

Then, problem (MP') becomes

$$\min_{0 \leq b_l \leq \min\{\pi - \alpha(l-1), \alpha\}} -b_l + \sin\frac{b_l}{2}$$

which is optimized when $b_l = \min\{\pi - \alpha(l-1), \alpha\}$. If $\pi - \alpha(l-1) < \alpha$, then the minimum candidate is

$$LC\left(\left\lceil\frac{\pi}{\alpha}\right\rceil\right) = 1 + \pi + 2\left(\left\lceil\frac{\pi}{\alpha}\right\rceil + 1\right)\sin\frac{\alpha}{2} \tag{14}$$

if $\frac{\pi}{\alpha} \notin \mathbb{N}$, and

$$LC\left(\frac{\pi}{\alpha}+1\right) = 1 + \pi + \left(\frac{2\pi}{\alpha}+4\right)\sin\frac{\alpha}{2} \qquad (15)$$

if $\frac{\pi}{\alpha} \in \mathbb{N}$. Since (15) is always greater or equal to the bound in (13), we will consider only the latter. If $\pi - \alpha(l-1) \geq \alpha$, then we have that $2 \leq l \leq \frac{\pi}{\alpha}$ and the minimum candidate is

$$LC\left(\left\lfloor\frac{\pi}{\alpha}\right\rfloor\right) = 1 + 2\pi - \alpha\left\lfloor\frac{\pi}{\alpha}\right\rfloor + 2\left(\left\lfloor\frac{\pi}{\alpha}\right\rfloor+1\right)\sin\frac{\alpha}{2} \qquad (16)$$

which holds only for $\alpha \leq \pi/2$. But (16) is lower than (7) for $\alpha \leq \pi/2$, and coincides with it for $\pi/2 \leq \alpha \leq \pi$. It is also lower than (15) for $\frac{\pi}{\alpha} \in \mathbb{N}$. Hence we will not be considering (7) at all.

The lower bound (for $\frac{\pi}{\alpha} \notin \mathbb{N}$) will be $\min\{LB(\lceil\frac{\pi}{\alpha}\rceil), LC(\lceil\frac{\pi}{\alpha}\rceil), LC(\lfloor\frac{\pi}{\alpha}\rfloor)\}$. But in this case, note that $LB(\lceil\frac{\pi}{\alpha}\rceil) \leq LC(\lceil\frac{\pi}{\alpha}\rceil) \ \forall\alpha$, so the lower bound will be $\min\{LB(\lceil\frac{\pi}{\alpha}\rceil), LC(\lfloor\frac{\pi}{\alpha}\rfloor)\}$.

$\square$

5   Qualitative Discussion About Our Results/Conclusions

We introduced a new optimization problem on *searching and fetching* with one robot from a unit disk, where there is limited information about the input. We studied two variants reflecting knowledge the robot has about its environment and contrasted how robot knowledge and capabilities affect the search time.

The goal of the current work was to provide nearly optimal algorithms with respect to worst case analysis. Indeed, in Sect. 3 we provided an optimal algorithm for problem 1-TE$_\geq$. Problem 1-TE$_=$ was studied in Sect. 4. The proposed algorithm is multiplicatively off from the optimal by at most the ratio between the expression derived in Theorem 2 (upper bound) over the expression derived in Theorem 3 (lower bound). It is not difficult to see that this value tends to 1 when α tends to either 0 or π, meaning that our algorithm is nearly optimal. Also, some tedious calculations show that this ratio is at most $\frac{4\sqrt{2}+3\pi+2}{6\sqrt{2}+2\pi+2} \approx 1.01868$ (and this value is attained when $\alpha = \pi/2$). In other words, our algorithm is provably off from the optimal by a multiplicative factor of at most 1.01868. This is

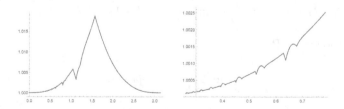

Fig. 4. The left-hand side plot is the ratio between the derived upper and lower bounds for problem 1-TE$_=$, for all values of α ranging from 0 to π. The right-hand side plot is a close-up for values of α between $\pi/12$ and $\pi/4$.

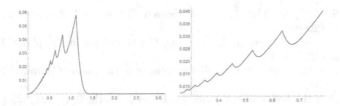

Fig. 5. The left-hand side plot is the difference between the naive Sweeping Algorithm and the derived upper bound for problem 1-TE$_=$, for all values of α ranging from 0 to π. The right-hand side plot is a close-up for values of α between $\pi/12$ and $\pi/4$.

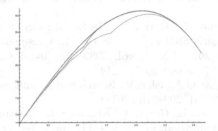

Fig. 6. Comparison between the cost of the naive sweeping algorithm (blue curve), the cost of Algorithm 2 (purple curve) and our lower bound (yellow curve). (Color figure online)

also summarized in Fig. 4. There is a lot of technical involvement for improving the performance of the Sweeping Algorithm for the problem. Figure 5 summarizes this improvement, while Fig. 6 shows all the upper and lower bounds in the discussion above. Notably, the improvement we obtain is linear in the radius of the disk when α is bounded away from 0 or π.

Given that our optimization problem has limited information about the input, it may be seen as an online task. Notably, our algorithms may not be nearly optimal with respect to competitive analysis. The optimal offline algorithm (knowing the exact locations of treasure and exit) has cost $1 + 2\sin(a/2)$. Comparing this to the performance of our online algorithm for problem 1-TE$_\geq$ gives rise to competitive ratio that is decreasing with α, and ranging from $1+2\pi$, when $\alpha = 0$, to $\frac{5+\pi}{3} \approx 2.71386$, when $\alpha = \pi$. For problem 1-TE$_=$, the induced competitive ratio exhibits very similar behavior. We leave it as an open problem as to whether the competitive ratio can be improved. That would be in addition to sharpening our bounds and to investigating our problem in other (continuous) geometric or discrete domains.

References

1. Ahlswede, R., Wegener, I.: Search Problems. Wiley-Interscience (1987)
2. Alpern, S.: Find-and-fetch search on a tree. Oper. Res. **59**(5), 1258–1268 (2011)

3. Alpern, S., Gal, S.: The Theory of Search Games and Rendezvous. Springer, Heidelberg (2003)
4. Baeza Yates, R., Culberson, J., Rawlins, G.: Searching in the plane. Inf. Comp. **106**(2), 234–252 (1993)
5. Baeza-Yates, R., Schott, R.: Parallel searching in the plane. Comput. Geom. **5**(3), 143–154 (1995)
6. Beck, A.: On the linear search problem. Israel J. Math. **2**(4), 221–228 (1964)
7. Bellman, R.: An optimal search. SIAM Rev. **5**(3), 274–274 (1963)
8. Berman, P.: On-line searching and navigation. In: Fiat, A., Woeginger, G.J. (eds.) Online Algorithms. LNCS, vol. 1442, pp. 232–241. Springer, Heidelberg (1998). doi:10.1007/BFb0029571
9. Bonato, A., Nowakowski, R.: The game of cops and robbers on graphs. In: AMS (2011)
10. Chrobak, M., Gąsieniec, L., Gorry, T., Martin, R.: Group search on the line. In: Italiano, G.F., Margaria-Steffen, T., Pokorný, J., Quisquater, J.-J., Wattenhofer, R. (eds.) SOFSEM 2015. LNCS, vol. 8939, pp. 164–176. Springer, Heidelberg (2015). doi:10.1007/978-3-662-46078-8_14
11. Chung, T.H., Hollinger, G.A., Isler, V.: Search and pursuit-evasion in mobile robotics. Auton. Robots **31**(4), 299–316 (2011)
12. Czyzowicz, J., Gąsieniec, L., Gorry, T., Kranakis, E., Martin, R., Pajak, D.: Evacuating robots via unknown exit in a disk. In: Kuhn, F. (ed.) DISC 2014. LNCS, vol. 8784, pp. 122–136. Springer, Heidelberg (2014). doi:10.1007/978-3-662-45174-8_9
13. Czyzowicz, J., Georgiou, K., Kranakis, E., Narayanan, L., Opatrny, J., Vogtenhuber, B.: Evacuating robots from a disk using face-to-face communication (extended abstract). In: Paschos, V.T., Widmayer, P. (eds.) CIAC 2015. LNCS, vol. 9079, pp. 140–152. Springer, Heidelberg (2015). doi:10.1007/978-3-319-18173-8_10
14. Czyzowicz, J., Kranakis, E., Krizanc, D., Narayanan, L., Opatrny, J., Shende, S.: Wireless autonomous robot evacuation from equilateral triangles and squares. In: ADHOC-NOW 2015, Athens, Greece, June 29–July 1, 2015, Proceedings, pp. 181–194 (2015)
15. Demaine, E.D., Fekete, S.P., Gal, S.: Online searching with turn cost. Theor. Comput. Sci. **361**(2), 342–355 (2006)
16. Fekete, S., Gray, C., Kröller, A.: Evacuation of rectilinear polygons. In: Wu, W., Daescu, O. (eds.) COCOA 2010. LNCS, vol. 6508, pp. 21–30. Springer, Heidelberg (2010). doi:10.1007/978-3-642-17458-2_3
17. Gluss, B.: An alternative solution to the lost at sea problem. Naval Res, Logistics Q. **8**(1), 117–122 (1961)
18. Isbell, J.R.: Pursuit around a hole. Naval Res. Logistics Q. **14**(4), 569–571 (1967)
19. Jennings, J.S., Whelan, G., Evans, W.F.: Cooperative search and rescue with a team of mobile robots. In: ICAR, pp. 193–200. IEEE (1997)
20. Kao, M.-Y., Ma, Y., Sipser, M., Yin, Y.: Optimal constructions of hybrid algorithms. J. Algorithms **29**(1), 142–164 (1998)
21. Kao, M.-Y., Reif, J.H., Tate, S.R.: Searching in an unknown environment: an optimal randomized algorithm for the cow-path problem. Inf. Comp. **131**(1), 63–79 (1996)
22. Koutsoupias, E., Papadimitriou, C., Yannakakis, M.: Searching a fixed graph. In: Meyer, F., Monien, B. (eds.) ICALP 1996. LNCS, vol. 1099, pp. 280–289. Springer, Heidelberg (1996). doi:10.1007/3-540-61440-0_135
23. Nahin, P., Chases, E.: The Mathematics of Pursuit and Evasion. Princeton University Press (2012)
24. Stone, L.: Theory of optimal search. Academic Press, New York (1975)

A 2-Approximation Algorithm for Barrier Coverage by Weighted Non-uniform Sensors on a Line

Robert Benkoczi, Daya Ram Gaur, and Mark Thom[(✉)]

Mathematics and Computer Science,
University of Lethbridge, Lethbridge, AB T1K 3M4, Canada
{benkoczi,gaur,thom}@cs.uleth.ca

Abstract. Barrier coverage is an approach to the intruder detection problem that relies on monitoring a perimeter, or barrier, of an area of interest using sensors placed around it. In this paper, we propose a weighted generalization of the unweighted line segment barrier coverage problem studied in [5] for which the authors demonstrate an FPTAS. We develop a fast, simple 2-approximation algorithm for the weighted case likely to be of interest to practitioners, and show that the FPTAS developed in [5] can be adapted to the weighted problem.

1 Introduction

Wireless sensor networks are used to provide detection services over an area of interest. Sensors are placed on the boundary, or perimeter, of the area. The perimeter to be covered is the *barrier*. If the sensors are placed such that any perimeter intrusion is sure to be detected, the sensors are said to cover the barrier. The covering of the perimeter by the sensors is also referred to as a *barrier coverage*, as the sensors are a barrier to the intruders.

In one-dimensional coverage problems, the barrier is represented by a horizontal line segment and sensors are initially placed on the line containing the line segment. The goal is to compute new positions for some subset of the given sensors, ensuring that every point in the barrier segment is within the detection range of some sensor. In typical two dimensional coverage problems, sensors are represented as points in the plane and assigned a radius of coverage while the barrier is represented as the boundary of some closed planar region.

Kumar *et al.* [11] were the first to formalize the barrier coverage by k sensors (k-barrier coverage). They established the equivalence between k-barrier coverage and the problem of determining k-vertex disjoint paths between a pair of

Wireless & Geometry Track

R. Benkoczi—Acknowledges the support for this research received from an NSERC Discovery Grant.

M. Thom—Acknowledges the support for this research received from the University of Lethbridge School of Graduate Studies.

© Springer International Publishing AG 2017
M. Chrobak et al. (Eds.): ALGOSENSORS 2016, LNCS 10050, pp. 95–111, 2017.
DOI: 10.1007/978-3-319-53058-1_7

vertices in a graph. The paper has spurred research on many aspects of barrier coverage problems, varying from density estimates of random deployments [2] to the relaxation of coverage requirements suited for the study of localized algorithms [6].

In a large scale deployment, using only the minimum number of sensors necessary to achieve full coverage is key to reducing costs. This discourages random deployments of sensors, as they are unlikely to make efficient use of detection ranges, resulting in few redundant sensors in the typical random covering.

In [8,9], Czyzowicz et al. introduced objective functions centered around the minimization of cost spiking quantities and studied the one dimensional barrier coverage problem. The MinMax problem minimizes the maximum distance travelled by any sensor; the MinNum problem minimizes the number of sensors used.

In [7], Chen et al. show the one-dimensional MinMax problem to be solvable in $O(n^2 \log n)$ time, where n is the number of sensors. It is possible to derive more efficient algorithms under the assumption that the sensing ranges are uniform. In [9], Czyzowicz et al. separate the barrier problem with uniform sensing ranges into two cases, those where the total length of sensing ranges exceeds the size of the line segment making up the barrier, and those where it does not. They provide exact algorithms for either case, with time complexity $O(n)$ for the second case and time complexity $O(n^2)$ for the first. Andrews and Wang [1] improved the $O(n^2)$ bound to $O(n \log n)$ for the case of uniform sensors. An extension of the line coverage problem where each point on the barrier is to be covered by k-sensors (k-line coverage) was studied recently by Wang et al. [12], who gave optimal algorithm for the case of uniform sensors.

Versions of two-dimensional MinMax and MinSum problems are addressed in [10], in which there are multiple barriers fixed as line segments in the plane. Sensors can have arbitrary sensing ranges and be located anywhere in the plane, and both Euclidean and rectilinear metrics of distance are considered. Variants of both problem types are introduced and distinguished according to the number of barriers to be covered. As well, the orientation of barriers plays a role, as barriers may be oriented parallel with or perpendicular to one another. While in general sensors may move freely, some variants require sensors to move to the closest point on a barrier. Dobrev et al. [10] demonstrate exact algorithms for MinMax and MinSum in the single barrier and k parallel barrier cases, and show that in all remaining cases, their variants of MinMax and MinSum are NP-hard.

Of chief interest to us here is the MinSum problem, which minimizes the total distance travelled by all sensors to their final positions in the solution. A common variation on the MinSum problem is the addition of some notion of sensor heterogeneity. In practice, some sensing units may consume less power or have greater sensing ranges, while newer units simply haven't had the time to suffer the effects of wear, deterioration and obsolescence. Therefore, barrier coverage models often allow for associated differences in the effectiveness of individual sensors to be expressed.

One model exhibiting sensor heterogeneity is found in Bar-Noy *et al.* [3]. They devise a sensing model where each sensor has finite battery life, and both moving and sensing cause sensors to consume battery power. Covering therefore occurs in two phases: the deployment phase, where sensors are moved into their final positions, and the covering phase, where sensors decide their sensing ranges to form a full barrier coverage. The amount of power consumed by movement and sensing is determined by given equations, which together establish the lifetime of each sensor. The goal is to maximize the barrier coverage lifetime, which is defined as the minimum lifetime of any sensor used in the coverage. For the case of sensors with variable sensing ranges, they [4] give an FPTAS to minimized the total energy, and an FPTAS to minimize the maximum energy spent. For the case of fixed sensing ranges they give an inapproximability result.

Benkoczi *et al.* [5] restrict the MinSum problem to the case where initially, sensors may only lie in positions where their detection ranges are disjoint from the barrier. Their model allows for sensing ranges to be non-uniform. They show the resulting *DisjointMinSum* problem to be NP-hard, and present a fully polynomial time approximation scheme (FPTAS) for *DisjointMinSum*, proving that solutions with approximation ratio arbitrarily close to 1 can always be computed for it in polynomial time. This result provides a new direction of investigation given that Czyzowicz *et al.* [8] proved that the unrestricted MinSum problem admits no constant factor approximation unless $P = NP$.

A generalization of DisjointMinSum unexplored in [5] is obtained by assigning individual weights to the movement costs of sensors, a second dimension of sensor heterogeneity after allowing ranges to be non-uniform. The objective of WeightedDisjointMinSum is to minimize the total weighted distance of sensors travelled in the solution under the constraints of DisjointMinSum.

In this paper, we develop a fast, simple 2-approximation algorithm for the WeightedDisjointMinSum problem. In cases where DisjointMinSum problems need to be solved quickly or using few computational resources, the speed and simplicity of the 2-approximation algorithm make it an attractive alternative to the FPTAS of [5], which contains a factor of $1/\epsilon$ in its running time.

This statement applies more generally to WeightedDisjointMinSum. By exploiting the existence of an ordering property previously shown to hold only in the unweighted case, we extend the results of [5] to generalize their FPTAS to the WeightedDisjointMinSum problem. The 2-approximation algorithm replaces the naive $n \cdot OPT$ approximation algorithm used in [5] to compute the movement budget in the LeftDisjointMinSum FPTAS, where n is the number of sensors in the problem instance and OPT is the cost of an optimal solution. As we will show, the substitution of a constant factor approximation algorithm in the computation of the budget amounts to a quadratic factor reduction in time complexity and a linear factor reduction in space complexity.

1.1 Notation and Problem Definitions

We are given a set of n wireless sensors S_i initially centered at points x_i. A sensor S_i has a detection range given by $r_i > 0$, triggering an alarm upon the detection of movement at any point x such that $d(x_i, x) \leq r_i$ where d is the Euclidean distance. We generalize the unweighted LeftDisjointMinSum problem as defined in [5] and adapt the notation used there.

The barrier is a line segment represented by the closed interval $I = [0, L]$ for some real $L > 0$. From its center point x_i, S_i detects movement in the range $I(S_i, x_i) = [x_i - r_i, x_i + r_i]$. After shifting S_i from its initial place m_i units, the new range of detection is $I(S_i, x_i + m_i) = [x_i - r_i + m_i, x_i + r_i + m_i]$.

We assume that $I(S_i, x_i) \cap I = \emptyset$, which is to say the range detected by each sensor is initially disjoint from I. For n sensors at center points $x_1 \leq x_2 \leq \ldots \leq x_n < 0$ where $x_i + r_i < 0$ for all $1 \leq i \leq n$, we define *WeightedLeftDisjointMinSum* as

$$\min \left\{ \sum_{1 \leq i \leq n} w_i |m_i| \right\} \text{ subject to } [0, L] \subseteq \bigcup_{i=1}^{n} I(S_i, x_i + m_i)$$

We appeal to the NP-hardness results of [5], where it is shown that LeftDisjointMinSum is NP-hard. Since LeftDisjointMinSum is a special case of WeightedLeftDisjointMinSum where $w_i = 1$ for all $1 \leq i \leq n$, it follows that WeightedLeftDisjointMinSum is NP-hard.

2   The 2-Approximation Algorithm

We present the algorithm in pseudocode, where S is the set of all weighted sensors, L is the length of the line segment to be covered and $d_i := |x_i| - r_i$ is the distance from the line segment to the outermost point of the detection range of s_i.

Algorithm 1. Calculate a covering of the line.

Require: $\sum_{s_i \in S} l_i \geq L$
   $Covering \leftarrow \emptyset$
   $x \leftarrow L$
   while $x > 0$ **do**
      $s \leftarrow \arg\min_{s_i \in S} \{ w_i(d_i + x) / \min(l_i, x) \}$
      $S \leftarrow S - \{s\}$
      $x \leftarrow x - l_s$
      $Covering \leftarrow Covering \cup \{s\}$
   end while
   return $Covering$

The algorithm is greedy, filling the line by packing sensors from the right. It selects one sensor per step, the minimum of the cost function $cost(s, x) := w_s(d_s + x) / \min(l_s, x)$ as it ranges over $s \in S$, with x fixed as the length of the prefix of the line that remains to be covered. The loop iterates until a full covering is produced.

We show that Algorithm 1 is a 2-approximation to WeightedLeftDisjoint-MinSum. The idea is to fix two solutions to the problem instance, the first being any covering of least cost, and the second the covering given by Algorithm 1. We iterate over the sensors of the approximate solution starting from the rightmost sensor and work our way leftward, accounting for the cost of each sensor in the approximate solution by no more than twice the cost of sensors in the optimal solution. We refer to the rightmost endpoint of the remaining succession of sensors, whose cost is still unaccounted for, in the approximate solution as the cursor.

We begin by decomposing the accounting procedure into lemmas handling various cases. Along the way, we identify invariants that must be preserved in each iterative step.

The first invariant describes what we term the *viability* of sensors in the optimal solution with respect to the approximate sensor at the cursor. When s was selected by Algorithm 1, it was determined to be the witness to the minimum of the function $s \to cost(s, x)$ over the set of candidate sensors for cursor x. The set of candidate sensors is identified as S in Algorithm 1's pseudocode.

Let S_x be the value of S at "time" x, and let app refer to the set of sensors comprising the solution given by Algorithm 1. Let opt be a fixed optimal solution. An optimal sensor $t \in S_x$ is said to be *viable* with respect to an approximate sensor s at x if $cost(s, x) \leq cost(t, x)$. The first invariant follows from the definition of Algorithm 1.

Invariant 1 (Viability). *Let* $s \in$ *app be the sensor at cursor* x, *so that* s *witnesses the minimum of the function* $s \to cost(s, x)$ *over* S_x. *If* $t \in opt \cap S_x$, *then* t *is viable with respect to* s *at* x.

Let $p_{s,T}$ be the location of the rightmost endpoint of sensor s in solution $T \in \{\text{app}, \text{opt}\}$. If $s \notin T$, we set $p_{s,T} = -\infty$. Define the *ratio* of sensor s as $ratio(s) := w_s/l_s$.

Lemma 1 (Order Preservation Property). *Let opt be an optimal solution, with sensors* $s, s' \in opt$ *such that* $p_{s,opt} \leq p_{s',opt}$. *Then*

$$ratio(s) \geq ratio(s')$$

Proof. By transitivity, we may suppose without loss of generality that s and s' are packed adjacently in opt, with $p_{s,\text{opt}} \leq p_{s',\text{opt}}$. Suppose for the sake of contradiction that

$$\frac{w_s}{l_s} < \frac{w_{s'}}{l_{s'}}$$

Which means that exchanging the positions of s and s' in opt yields a change in cost

$$w_s l_{s'} - w_{s'} l_s < 0$$

This is a contradiction. □

Three technical results we will often use are given in the following lemmas.

Lemma 2 (Invariance of Cost and Ratio under Fractioning). *Let $s \in opt$ and let the number $l_{s'}$ satisfy $0 < l_{s'} \le l_s$. Designate a contiguous length of s as sensor s' of length $l_{s'}$ and weight $w_{s'} = (l_{s'}/l_s)w_s$ (we refer to this procedure as fractioning). Then $cost(s,x) = cost(s',x)$ and $ratio(s) = ratio(s')$ for all $x > l_{s'}$.*

Proof. A matter of expanding definitions and doing arithmetic. □

Fact 2. *Let $a_1, \ldots, a_n$, $b_1, \ldots, b_n$ be nonnegative real numbers, $b_i \ne 0$. Then*

$$\min_{1 \le i \le n} \frac{a_i}{b_i} \le \frac{a_1 + \ldots + a_n}{b_1 + \ldots + b_n} \le \max_{1 \le i \le n} \frac{a_i}{b_i}$$

Lemma 3 (Shifting trick). *Let $s \in opt$, and for $1 \le i \le n$, $s_i \in opt$. Suppose $p_{s,opt} \le p_{s_i,opt}$ for each $1 \le i \le n$ and $\sum_{i=1}^{n} l_{s_i} \le l_s$. Then*

$$w_s \ge \sum_{i=1}^{n} w_{s_i}$$

Proof. From Lemma 1 we have $w_s/l_s \ge w_{s_i}/l_{s_i}$ for each $1 \le i \le n$. From the righthand inequality of Fact 2 we get

$$\frac{w_s}{l_s} \ge \max_{1 \le i \le n} \frac{w_{s_i}}{l_{s_i}} \ge \frac{\sum_{i=1}^{n} w_{s_i}}{\sum_{i=1}^{n} l_{s_i}}$$

Along with $\sum_{i=1}^{n} l_{s_i} \le l_s$, we may conclude that $w_s \ge \sum_{i=1}^{n} w_{s_i}$. □

With the lemmas in hand, we develop the accounting procedure.

Proposition 1. *Suppose Invariant 1 holds and that $s \in opt$ is the sole element of opt, meaning that s is at the cursor x, and $l_s \ge x$. Then the total cost of sensors $s' \in app$ is no more than twice the cost of s.*

Proof. Let $s_n \in app$ be the sensor at cursor x and $s_1, \ldots, s_{n-1}$ the remaining sensors in app in order starting from the left. We assume that the left end of the barrier, point 0, is covered by sensor s_1.

Since s_1 is needed to cover the left end point, we must have $\sum_{i=2}^{n} l_{s_i} < x \le l_s$. $l_s \ge x$ implies $s \ne s_i$ for each $2 \le i \le n$, and thus, $s \in S_{x-\sum_{i=j+1}^{n} l_{s_i}}$ for each $2 \le j \le n$, and so is viable with respect to each s_j at $x - \sum_{i=j+1}^{n} l_{s_i}$.

From this observation and Fact 2, we derive

$$\frac{w_s(d_s + x)}{x} \ge \max_{2 \le i \le n} \frac{w_{s_i}(d_{s_i} + x - \sum_{j=i+1}^{n} l_{s_j})}{l_{s_i}} \ge \frac{\sum_{i=2}^{n} w_{s_i}(d_{s_i} + x - \sum_{j=i+1}^{n} l_{s_j})}{\sum_{i=2}^{n} l_{s_i}}$$

Since $x \ge \sum_{i=2}^{n} l_{s_i}$, it is clear that $w_s(d_s + x) \ge \sum_{i=2}^{n} w_{s_i}(d_{s_i} + x - \sum_{j=i+1}^{n} l_{s_j})$.

Now, if $s = s_1$, then the cost of s_1 can be paid using a second charge to the cost of s. If $s \ne s_1$, then s is viable with respect to s_1 at $p_{s_1,app}$ and a similar bounding argument shows we can pay for s_1 using a second charge to the cost of s. □

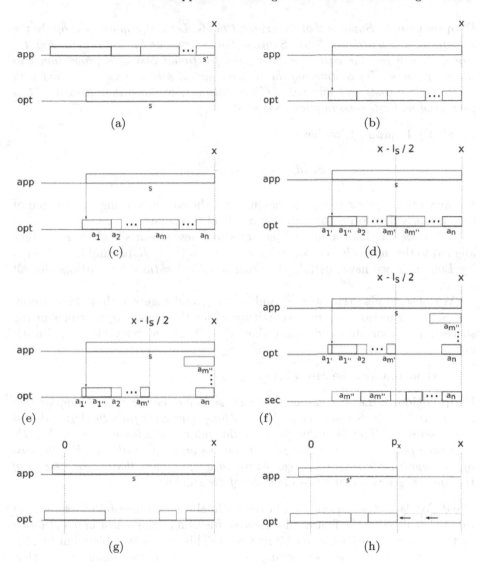

Fig. 1. (a) A single sensor covering. (b) A multiple sensor covering. (c) Stabbing a_m. (d) Compaction and the secondary line. (e) A single sensor s in app with gaps. (f) s is shrunk to s'.

We will assume from this point forward that $s \in$ opt at the cursor x has its full length contained within the covering interval $[0, L]$, which is to say that

$$cost(s, x) = \frac{w_s(d_s + x)}{\min(l_s, x)} = \frac{w_s(d_s + x)}{l_s}$$

If this assumption does not hold, we will apply a simple adaptation of Proposition 1.

Proposition 2. *Suppose that Invariant 1 holds. Let $s' \in app$ and $s \in opt$ be the sensors at x, such that $s \in S_x$. Suppose that $x \geq l_{s'}$ and $l_s > l_{s'}$ as in Fig. 1a. Then we can pay the cost of s' using the fractional cost of s proportional to the length of s'. By contracting the unused part of s to a sensor $s'' \in opt$ with $cost(s'', x) = cost(s, x)$ and $ratio(s'') = ratio(s)$, we have that Invariant 1 is preserved under the assumption that $s'' \in S_{x-l_{s'}}$.*

Proof. By Invariant 1, we have

$$w_{s'}(d_{s'} + x) \leq l_{s'} \frac{w_s(d_s + x)}{l_s}$$

We move the cursor x to $x - l_{s'}$, having paid the cost of s' using the fraction of the cost of s that forms the right-hand side of the bound.

Then, we fraction the unused part of s to a new sensor s'', where s'' is right aligned to the succeeding cursor $x - l_{s'}$ with $w_{s''} = (1 - l_{s'}/l_s)w_s$ and $l_{s''} = l_s - l_{s'}$. By Lemma 2, we have $cost(s'', y) = cost(s, y)$ and $ratio(s'') = ratio(s)$ for all $y > 0$.

We observe also that $s \neq s'$ and $s \in S_x$, and together these facts imply $s \in S_{x-l_{s'}}$. Therefore, we may substitute s'' for the remaining part of s in the succeeding cursor step and assert that $s'' \in S_{x-l_{s'}}$ without violating Invariant 1. □

Next we consider the case where $l_{s'} > l_s$.

Proposition 3. *Let s be the approximate sensor at cursor x, and suppose that Invariant 1 holds. Suppose we find by stabbing downward from the left endpoint of s a sensor distinct from that found in the optimal solution at x, as in Fig. 1b, with each sensor on and to the right of stabline an element of S_x. Then the cost of the approximate sensor at x can be bounded by no more than twice the cost of the optimal sensors that lie to the right of the stabline.*

Proof. We label the sensors used in the optimal solution for ease of reference in Fig. 1c. As noted in the figure, s is bisected, fractioning the sensor in opt stabbed by the bisection line (below, a_m) if necessary. This process is depicted in Fig. 1d.

Above, the optimal sensor a_m is fractioned at the bisector, so that $w_{a'_m} = (l_{a'_m}/l_{a_m})w_{a_m}$ and $w_{a''_m} = (l_{a''_m}/l_{a_m})w_{a_m}$. By Lemma 2, $ratio(a'_m) = ratio(a_{m''}) = ratio(a_m)$ and $cost(a_{m'}, y) = cost(a_{m''}, y) = cost(a_m, y)$ for all $y > 0$.

Next, we compact the optimal sensors lying to the right of the bisector to right-align with x, as in Fig. 1e.

The extra charge needed to perform the compaction is not included in the costs tallied in the optimal solution. In order to pay for the charge, we charge each sensor $a_{m''}, \ldots, a_n$ twice, and deduct the charge from the secondary charge of each compacted sensor, as depicted below on the *sec* line. See Fig. 1f.

The dashed sensors in *sec* correspond to empty space of length equal to the length of the labelled sensor. Of the compacted sensors, $a_{m''}$ is furthest to the left, and the charge required to compact it is equal to the length of all the sensors following it in opt. In sec, $a_{m''}$ is deducted the same total length, retracted back from its original position at $p_{a_{m''},sec} = p_{a_{m''},opt}$. This is to compensate for the charge added to it in opt, since we want to charge $a_{m''}$ no more than twice its original cost. The same compensation is applied to the charge required by each of the compacted sensors, leaving the solid sensors in sec to represent the unused portions of the secondary charges.

Since $a_{m''}, \ldots, a_n \in S_x$, we have by Fact 2

$$\frac{\sum_{i=m''}^{n} w_{a_i}(d_{a_i} + x)}{\sum_{i=m''}^{n} l_{a_i}} \geq \frac{w_s(d_s + x)}{l_s}$$

Because $\sum_{i=m''}^{n} l_{a_i} = l_s/2$, we have $\sum_{i=m''}^{n} w_{a_i}(d_{a_i} + x) \geq w_s(d_s + x)/2$. To pay for the second half of $w_s(d_s + x)$, we have to use the remaining sensors to the left of the bisector alongside the remains of the secondary charges. We shift each of the solid sensors in sec to align with x by using the untapped charge of the sensors to the left of the bisector in opt, as follows.

Starting from $x - l_s/2$, we trawl leftward, capturing the sensors comprising the interval $I_{a_{m''}} = [x - l_s/2 - l_{a_{m''}}, x - l_s/2]$, fractionally if necessary. If sensor $a \in I_{a_{m''}}$, then by Lemma 1 we have $ratio(a) \geq ratio(a_{m''})$. Since $\sum_{a \in I_{a_{m''}}} l_a = l_{a_{m''}}$, we can apply Lemma 3 to get $\sum_{a \in I_{a_{m''}}} w_a \geq w_{a_{m''}}$.

We note that for every point $p \in I_{a_{m''}}$, $p - (x - l_s) \geq l_s/2 - l_{a_{m''}}$. By charging every interval in $I_{a_{m''}}$ twice, we get $(l_s - 2l_{a_{m''}})(\sum_{a \in I_{a_{m''}}} w_a)$ in charge. By the inequality just obtained, we have

$$(l_s - 2l_{a_{m''}}) \sum_{a \in I_{a_{m''}}} w_a \geq (l_s - 2l_{a_{m''}})w_{a_{m''}}$$

which is the minimum charge necessary to shift $a_{m''} \in sec$ to right-align with x.

The same argument serves to move the remaining sensors in opt to right align with x, with the lengths adjusted to match the distance of each sensor in sec from x. Once each copy of $a_{m''}, \ldots, a_n$ is aligned to x, we pay for the second half of $w_s(d_s + x)$ as before. □

It should also be noted that the fractional part of a_1 unused in the proof of Proposition 3, $a_{1'}$, can be supposed to be an element of S_{x-l_s}. This follows by an argument similar to the one used to prove the analogous statement at the end of Proposition 2.

We might be tempted to adapt the argument of Proposition 3 to an entire approximate solution by bisecting the whole of the optimal solution and applying the charging scheme. Since the proof of Proposition 3 depends on every optimal sensor in the range underlying s at cursor x being a member of S_x, the scheme cannot work in every case. We could easily conceive of a weighted MinSum problem in which $p_{t,opt} = p_{t,app} = L$ and $t \notin S_x$ for any $x < t$, leaving the second copy of t unused to charge for any part of opt except t at L.

Since the cases we now have all depend on the condition of some optimal sensors near x being in S_x, the only cases left to consider are those in which the condition fails for optimal sensors near x. An optimal sensor $t \notin S_x$ if and only if t appeared in app after s. Equivalently, t was choosen by the algorithm before s.

Therefore, we give special consideration to sensors $t \in$ app $\cap$ opt. Namely, we ask whether charging t twice is enough to shift t to x. If it is, we simply shift t to the cursor by charging t twice, pay for $t \in$ app using the shifted $t \in$ opt, compact the gap left by $t \in$ opt, and resume the procedure at the next approximate sensor.

If it is not, we use the shifting trick as follows. The charge of t is used to move l_t in total length of sensors u in the interval $[x - p_{t,\text{opt}}, x] \cap$ opt to x. We note that no part of t intersects $[x - p_{t,\text{opt}}, x]$ in opt. Then, shifted sensors $u \in$ opt are greedily selected in descending order of $p_{u,\text{app}}$. Since $p_{u,\text{app}} = -\infty$ for all $u \in$ opt $-$ app, this is to say we will always prefer $u \in$ app over $u \notin$ app.

All sensors u selected for shifting are viable with respect to s at x, since for each u we have $p_{u,\text{app}} \leq p_{s,\text{app}}$, thereby implying $u \in S_x$. $s \in$ app is paid using the sensors u, the gaps left by the sensors u are compacted, and the gap left by t is kept in opt. This approach is technically detailed and verified in the following two propositions. Their proofs can be found in the appendix.

Proposition 4. *Suppose the approximate sensor s at cursor x satisfies $2p_{s,\text{opt}} \geq x$. Then the cost of $s \in$ app can be paid entirely by the cost of $s \in$ opt.*

Proposition 5. *Suppose the approximate sensor s at cursor x satisfies $2p_{s,\text{opt}} < x$ and that the subinterval of sensor placements $[x - p_{s,\text{opt}}, x] \subseteq$ opt contains no gaps. Then the cost of $s \in$ app can be paid using the cost of $s \in$ opt, and a length of l_s in sensors belonging to $[x - p_{s,\text{opt}}, x] \subseteq$ opt.*

We have left open the problem of how to address the presence of gaps in opt. We especially want to guard against the possibility of gaps showing up near x, which could preclude the use of earlier propositions depending on a sufficient length of optimal sensors being near x.

To deal with gaps, we note that in the summary of the previous two propositions, we charged the costs of t and each of the shifted sensors u only once. We could instead charge t twice, to get two copies of each u shifted to x. As before, the first copy of each u is used to pay for s whereas the second copy of each u is cached for later use.

We introduce an invariant relating the total size of the gaps in opt and the total size of sensors u stored in the cache. The cache is actually a priority queue. The order of the elements in the priority queue is the same order that is used to select the shifted u.

Invariant 3. *The total length of sensors in the priority queue PQ is equal to the total size of gaps in opt. Furthermore, for every $t \in PQ$ we have $p_{t,\text{opt}} \leq x$ where x is the cursor.*

If the priority queue is non-empty and the approximate sensor s at the cursor is not in opt, we draw sensors from the priority queue to pay for s. By compacting the gaps closest to the cursor by the length of sensors extracted from the queue, Invariant 3 is preserved.

If s is in opt and the conditions of Proposition 4 apply, then s is paid using Proposition 4. Otherwise, we want to use the shifting trick again, so that $s \in$ opt never appears near x.

In order for the shifting trick to apply, we must ensure that some contiguous range of viable sensors of total length at least l_s is aligned at x. To that end, we introduce a final invariant.

For every gap g in opt, we maintain a gapless "active region" r_g of size large enough that the shifting trick has an adequately sized pool of sensors from which to draw. Every cached sensor t added to the priority queue after the formation of gap g will be tagged as belonging to g, which we denote as $tag(t) = g$.

Thus, every sensor in the priority queue is tagged as belonging to a unique gap g, and g is closed at the exact time the last sensor tagged g is removed from the priority queue.

Invariant 4 (Total size of active region with respect to gaps). *Let g be a gap in opt. Then there is a contiguous (gapless) region r_g right-aligned at the cursor in opt, whose length is equal to the total size of sensors lying to the left of g in opt. Furthermore, if sensor $s \in r_g$ we have $p_{s,\mathrm{opt}} \leq p_g$, where $p_g = \min\{p_{t,\mathrm{opt}} : t \in PQ, tag(t) = g\}$.*

For $s \in r_g$, the last statement relating $p_{s,\mathrm{opt}}$ to p_g ensures that sensors $t \in$ app $\cap$ opt of the type described in Proposition 4 fall outside of r_g, and thus do not diminish the size of r_g when $s \in$ opt is spent and substituted with a gap. To maintain Invariant 4, we have to be more careful when selecting shifting sensors t in Proposition 5. The technical details are explained in the proof of Proposition 6, which ties the structure of the bounding together using the earlier propositions to dispatch the various subcases.

Since we can now replace sensors with gaps anywhere in the contiguous sensor lines of opt, we should generalize the concept of a sensor line to that of a *sensor configuration*, which is otherwise the same but can include gaps and left-shifted sensors. That is, a sensor configuration T_C is a subset of the sensor line $T \in$ {app, opt} such that for each $t \in T_C$, $p_{t,C} \leq p_{t,T}$. We note that sensors in app$_C$ are never shifted, and are only ever eliminated at the cursor, so that app$_C$ never contains gaps between sensors. The proof of Proposition 6 is contained in the appendix.

Proposition 6. *(Bounding procedure) Suppose we have sensor configurations app$_C$ and opt$_C$ satisfying the conditions of Invariants 1, 3 and 4. Then the total cost of sensors in app$_C$ is no more than twice the total cost of the sensors in opt$_C$ plus the cost of sensors in the priority queue.*

Theorem 5 (Algorithm 1 is a 2-approximation to WeightedMinSum). *Let app be a solution to an instance of the WeightedMinSum problem with cost APP and let opt be any fixed optimal solution for the same instance with cost OPT. Then APP $\leq$ 2OPT.*

Proof. We invoke Proposition 6, noting that all three Invariants are trivially satisfied by the solution computed by Algorithm 1.   □

Theorem 6 (Tightness of the approximation bound). *The 2-approximation bound for Algorithm 1 is tight.*

Proof. We construct a parametric family of covering problems whose solution costs under Algorithm 1 are arbitrarily close to twice their optimal solution costs.

Let $L > 0$ be the length of the line to be covered. Lying disjointly to the left of the line are two sensors. The first sensor, s_1, has length $l_1 = cL$ and distance from the line $d_1 = 1 - \epsilon$ for some $\epsilon > 0$ and $1 - \epsilon < c < 1$, with weight $w_1 = c$.

Similarly, sensor s_2 has $l_2 = L$, $d_2 = c$ and $w_2 = 1$. It is readily seen that the optimal solution consists only of s_2, so that $OPT = c + L$.

Running Algorithm 1 on the instance produces an ordered solution of $< s_2, s_1 >$, with cost $c(1 - \epsilon) + cL + c + L - cL = c(1 - \epsilon) + c + L$, giving the approximation ratio

$$\frac{c(1 - \epsilon) + c + L}{c + L} = \frac{1 - \epsilon}{1 + L/c} + 1$$

Let $L = c\epsilon$. Since we have free choice of $\epsilon > 0$ and $\lim_{\epsilon \to 0^+} (1 - \epsilon)/(1 + \epsilon) = 1$, the approximation bound for the instance can come as close to 2 as we wish, demonstrating that the bound is tight. $\qquad\square$

3   FPTAS for WeightedDisjointMinSum

See Sect. B in the Appendix for a discussion of the FPTAS for the weighted barrier coverage problem.

4   Conclusion

We have shown a simple, efficient 2-approximation algorithm for the WeightedLeftDisjointMinSum problem. Along with the Order Preservation Property, we used it to generalize the unweighted DisjointMinSum FPTAS of [5] to the WeightedDisjointMinSum problem. DisjointMinSum is a generalization of LeftDisjointMinSum that allows disjointly ranged sensors to fall on either side of the line segment barrier. By replacing the naive $n \cdot OPT$ approximation algorithm used in the derivation of the FPTAS in [5] with our worst-case quadratic 2-approximation algorithm, we obtain a quadratic factor reduction in the time complexity of the WeightedDisjointMinSum FPTAS and a linear factor reduction in its space complexity.

A   Appendix

A.1   Proofs from Sect. 3

Proof (of Proposition 4). We charge $s \in$ opt twice, noting

$$2w_s(d_s + p_{s,\text{opt}}) \geq w_s(d_s + 2p_{s,\text{opt}}) \geq w_s(d_s + x)$$

$\qquad\square$

Proof (of Proposition 5). We break the proof into cases centered around the total size of the sensors in the set $\mathcal{I} = \{t \in \text{app} : t \in [x - p_{s,\text{opt}}, x] \cap \text{opt}\}$.

If $|\mathcal{I}|_{size} = \sum_{t \in \mathcal{I}} l_t \geq l_s$, we select l_s in total length of $t \in \mathcal{I}$ and in order of descending $p_{t,\text{opt}}$. Let $\mathcal{I}'$ be this subset of $\mathcal{I}$, so that $|\mathcal{I}'|_{size} = l_s$. Since $2p_{s,\text{opt}} < x$ and $t \in \mathcal{I}$ implies $t \in [x - p_{s,\text{opt}}, x]$, we have $p_{s,\text{opt}} < p_{t,\text{opt}}$ for all $t \in \mathcal{I}$. Therefore, $ratio(s) \geq ratio(t)$ for all $t \in \mathcal{I}$ by Lemma 1.

From Fact 2, we have

$$\frac{w_s}{l_s} \geq \frac{\sum_{t \in \mathcal{I}'} w_t}{\sum_{t \in \mathcal{I}'} l_t} \geq \frac{\sum_{t \in \mathcal{I}'} w_t}{l_s}$$

so that $w_s \geq \sum_{t \in \mathcal{I}'} w_t$.

Since $|x - p_{t,\text{opt}}| \leq p_{s,\text{opt}}$ for each $t \in \mathcal{I}'$, we can use a single charge of s to move each $t \in \mathcal{I}'$ to right-align with x. We notice that each $t \in \mathcal{I}$ is viable with respect to s at x, since t precedes s in app. Since $|\mathcal{I}'|_{size} = l_s$, we can pay for $s \in$ app using a single charge of $s \in$ opt and a single charge of each $t \in \mathcal{I}'$. The process is concluded by substituting a gap for $s \in$ opt and for each $t \in \mathcal{I}'$.

We note that this is an adaptation of the shifting trick originally developed for use in Proposition 3. If instead $|\mathcal{I}|_{size} < l_s$, we use the same shifting trick to move each $t \in \mathcal{I}$ to right-align with x, charging the additional charge to $s \in$ opt. This leaves s with $l_s - |\mathcal{I}|_{size}$ of a single charge remaining, which we use to shift $l_s - |\mathcal{I}|_{size}$ in total of sensors (any sensors) in $[x - p_{s,\text{opt}}, x] - \mathcal{I}$.

The sensors we select to shift, whether in $\mathcal{I}$ or not, are viable with respect to $s \in$ app at x. The selected sensors not in $\mathcal{I}$ aren't found in app, and so, are elements of S_x and therefore viable with respect to s at x. Once more, the cost of s is paid and the selected sensors moved by s and $s \in$ opt itself are replaced by gaps once their charges are spent. □

Proof (of Proposition 6). By induction on the number of sensors whose cost remains to be bounded in app_C.

Suppose $n = 1$. A single sensor is in app_C, with the possibility of sensors punctuated by gaps "below" it in opt_C, as depicted in Fig. 1g.

By Invariants 1 and 3, we can pay for exactly the proportion of s equal to the total length of the gaps using the cost of sensors drawn from the priority queue. If this does not cover the cost of s entirely, the configuration is made to resemble Fig. 1h, where s is shrunk down to s', the remaining unpaid length of s, with the sensors in opt_C compacted to align with the new cursor p_x.

Since each $t \in \text{opt}_C$ is distinct from s (except possibly the t containing 0, but this makes no difference to the argument), we have $t \in S_x$. By Invariant 1, the sensors $t \in \text{opt}_C$ satisfy

$$\frac{w_s(d_s + x)}{\min(x, l_s)} \leq \frac{w_t(d_t + x)}{\min(x, l_t)}$$

and therefore, with $p = p_x/x$ the unpaid portion of s,

$$p \cdot \frac{w_s(d_s + x)}{\min(x, l_s)} \leq p \cdot \frac{w_t(d_t + x)}{\min(x, l_t)}$$
$$\leq \frac{w_t(d_t + p \cdot x)}{\min(x, l_t)}$$
$$\leq \frac{w_t(d_t + p_x)}{\min(x, l_t)}$$

With the conditions of Propositions 1, 2 and 3 all satisfied, we conclude the case with whichever of the three is applicable.

Now suppose $n = k + 1$ and that the statement of the Proposition holds if app_C has k or fewer sensors. We consider the sensor s at cursor x by case.

Case 1: $s \in \mathrm{app}_C \cap \mathrm{opt}_C$

If $2p_{s,\mathrm{opt}_C} \geq x$, we apply Proposition 4 to pay for $s \in \mathrm{app}_C$, closing the gap left in place of s by shifting the optimal sensors to the right of s leftward. We note that Invariants 1, 3 and 4 are maintained after the shift.

If $2p_{s,\mathrm{opt}_C} < x$, we apply Proposition 5 with a modification to its argument made to maintain Invariant 4. Let $u_1, \ldots u_n \in \mathrm{opt}_C$ be the first l_s in total length of sensors u satisfying $2p_{u,\mathrm{opt}_C} < p_{u,\mathrm{app}_C}$ leftward from x. We use the sensors u_i instead of s in providing the charge to shift the selected sensors in the range $[x - p_{s,\mathrm{opt}_C}, x]$ forward. Gaps $g_{u_1}, \ldots, g_{u_n}$ are introduced in their place, and s is carved into contiguous fragments of length $l_{u_1}, \ldots, l_{u_n}$, which are then identified with sensors $u_1, \ldots, u_n \in \mathrm{app}_C$ respectively.

Since $2p_{s,\mathrm{opt}_C} \leq 2p_{u_i,\mathrm{opt}_c} < p_{u_i,\mathrm{app}_c}$ for each u_i, we can consider that of all the other optimal sensors of its type, s was nearest to x all along, although that may not have been true. Under this assumption, it's clear that Invariant 4 is maintained, since any gaps following s in opt_C correspond to larger gap regions, and the secondary sensors t selected in those cases therefore have larger p_{t,opt_C} values.

Secondary copies of $u_1, \ldots, u_n$ are added to the priority queue, and as we've just opened gaps of those sizes, Invariant 3 is preserved. It is also clear that Invariant 1 is preserved.

Case 2: $s \notin \mathrm{app}_C \cap \mathrm{opt}_C$.

If the priority queue is not empty, we use Invariant 3 to draw viable sensors from the priority queue to pay for s.

Suppose we can entirely pay for s using auxiliary sensors drawn from the queue. Since $p_{t,\mathrm{opt}_C} \leq x$ for all $t \in PQ$, and t is ordered maximally by p_{t,opt_C}, we have that the sensors t' remaining in the queue after s is paid must satisfy $p_{t',\mathrm{opt}_C} \leq x - l_s$, since every positive p_{t,opt_c} is unique. If the auxiliary sensors in the priority queue can only pay for a partial length of s (or no length, which is to say the queue is empty), we fall back on the argument given in case $n = 1$, which does not rely on the assumption that s is the only sensor left in app_C.

In either case, Invariant 3 is preserved and the gaps tagged by the secondary sensors are compacted against by the length of those sensors. The remaining regions and gaps are unaffected except as a result of compaction. Because of the queue ordering, the only effect this has is to reduce or close the rightmost gaps, leaving Invariant 4 intact.   □

B   The WeightedDisjointMinSum FPTAS

The FPTAS of [5] approximates the LeftDisjointMinSum problem, and is derived from a discretization of the continuous recurrence given by

$$f^*(i+1, z) = \max\{\min\{f^*(i, z), g^*(i+1, z)\}, 0\}$$

where

$$g^*(i+1, z) = \min_{x \in [0,z]} \{f^*(i, z-x) - 2r_{i+1} : |f^*(i, z-x) - r_{i+1} - x_{i+1}| \le x\}$$

and

$$f^*(0, z) = 0 \text{ for all } z \in [0, +\infty)$$

OPT is computed as

$$OPT = \min_{z \ge 0}\{z : f^*(n, z) \le 0\}$$

Aside from the use of z to denote the budget, the notation used here matches our own. First, the sensors are ordered by Lemma 1, each of them of unit weight. $f^*(i, z)$ is the length of the largest subcovering $[f^*(i, z), L]$ formed from the first i sensors whose total movement cost is no greater than z.

We see that the recurrence can use sensor $i + 1$ to augment the best possible coverage of i sensors using budget $z - x$ for any $0 \le x \le z$, where x is the part of the budget funding the movement of sensor $(i + 1)$. The length of the old covering is extended by $2r_{i+1}$ units if $|f^*(i, z-x) - r_{i+1} - x_{i+1}| \le x$, meaning that x is enough to pay for the movement of sensor $(i + 1)$. If sensor $(i + 1)$ is not used, the recurrence reverts to the optimal covering under z restricted to the first i sensors, of cost $f^*(i, z)$.

Sorting the sensors as prescribed by Lemma 1, we rewrite the continuous recurrence for WeightedLeftDisjointMinSum as

$$f^*(i+1, z) = \max\{\min\{f^*(i, z), g^*(i+1, z)\}, 0\}$$

where

$$g^*(i+1, z) = \min_{x \in [0,z]} \{f^*(i, z-x) - 2r_{i+1} : w_{i+1} \cdot |f^*(i, z-x) - r_{i+1} - x_{i+1}| \le x\}$$

and

$$f^*(0, z) = 0 \text{ for all } z \in [0, +\infty)$$

The second dimension of the continuous dynamic programming table is discretized in units of $\zeta = \epsilon Z/(n(n+1))$ over the interval $[0, Z]$, where Z is the upper bound on budgets. Z is taken to be the cost of the solution produced by the naive $n \cdot OPT$ approximation algorithm described in [5], meaning $Z \leq n \cdot OPT$.

Substituting our 2-approximation algorithm for the naive $n \cdot OPT$ algorithm we revise the bound to $Z \leq 2 \cdot OPT$. ζ becomes $\zeta = \epsilon Z/2(n+1)$, so that the units of ζ that form the second dimension of the discretized table are between 0 and $Z/\zeta + n + 1 = 2(n+1)/\epsilon + n + 1$. With these change in place, the derivation and proof of correctness of the FPTAS for WeightedDisjointMinSum proceeds as it does for DisjointMinSum in [5].

We gain a quadratic factor improvement in run time, since the work of computing g drops from quadratic to linear time, resulting in time complexity $O(n^5/\epsilon^3)$ for the whole FPTAS.

References

1. Andrews, A.M., Wang, H.: Minimizing the aggregate movements for interval coverage. In: Dehne, F., Sack, J.-R., Stege, U. (eds.) WADS 2015. LNCS, vol. 9214, pp. 28–39. Springer, Heidelberg (2015). doi:10.1007/978-3-319-21840-3_3
2. Balister, P., Bollobas, B., Sarkar, A., Kumar, S.: Reliable density estimates for coverage and connectivity in thin strips of finite length. In: Proceedings of the 13th Annual ACM International Conference on Mobile Computing and Networking, MobiCom 2007, pp. 75–86. ACM, New York (2007)
3. Bar-Noy, A., Rawitz, D., Terlecky, P.: Maximizing barrier coverage lifetime with mobile sensors. In: Bodlaender, H.L., Italiano, G.F. (eds.) ESA 2013. LNCS, vol. 8125, pp. 97–108. Springer, Heidelberg (2013). doi:10.1007/978-3-642-40450-4_9
4. Bar-Noy, A., Rawitz, D., Terlecky, P.: "Green" barrier coverage with mobile sensors. In: Paschos, V.T., Widmayer, P. (eds.) CIAC 2015. LNCS, vol. 9079, pp. 33–46. Springer, Heidelberg (2015). doi:10.1007/978-3-319-18173-8_2
5. Benkoczi, R., Friggstad, Z., Gaur, D., Thom, M.: Minimizing total sensor movement for barrier coverage by non-uniform sensors on a line. In: Bose, P., Gąsieniec, L.A., Römer, K., Wattenhofer, R. (eds.) ALGOSENSORS 2015. LNCS, vol. 9536, pp. 98–111. Springer, Heidelberg (2015). doi:10.1007/978-3-319-28472-9_8
6. Chen, A., Kumar, S., Lai, T.H.: Designing localized algorithms for barrier coverage. In: Proceedings of the 13th Annual ACM International Conference on Mobile Computing and Networking, MobiCom 2007, pp. 63–74. ACM, New York (2007)
7. Chen, D.Z., Gu, Y., Li, J., Wang, H.: Algorithms on minimizing the maximum sensor movement for barrier coverage of a linear domain. Discrete Comput. Geometry 50(2), 374–408 (2013)
8. Czyzowicz, J., Kranakis, E., Krizanc, D., Lambadaris, I., Narayanan, L., Opatrny, J., Stacho, L., Urrutia, J., Yazdani, M.: On minimizing the sum of sensor movements for barrier coverage of a line segment. In: Nikolaidis, I., Wu, K. (eds.) ADHOC-NOW 2010. LNCS, vol. 6288, pp. 29–42. Springer, Heidelberg (2010). doi:10.1007/978-3-642-14785-2_3
9. Czyzowicz, J., Kranakis, E., Krizanc, D., Lambadaris, I., Narayanan, L., Stacho, L., Urrutia, J., Yazdani, M.: On minimizing the maximum sensor movement for barrier coverage of a line segment. In: Proceedings of 8th International Conference on Ad Hoc Networks and Wireless, pp. 22–25 (2002)

10. Dobrev, S., Durocher, S., Eftekhari, M., Georgiou, K., Kranakis, E., Krizanc, D., Narayanan, L., Opatrny, J., Shende, S., Urrutia, J.: Complexity of barrier coverage with relocatable sensors in the plane. Theoret. Comput. Sci. **579**, 64–73 (2015)
11. Kumar, S., Lai, T.H., Arora, A.: Barrier coverage with wireless sensors. In: Proceedings of the 11th Annual International Conference on Mobile Computing and Networking, pp. 284–298. ACM (2005)
12. Wang, Y., Wu, S., Gao, X., Wu, F., Chen, G.: Minimizing mobile sensor movements to form a line k-coverage. Peer-to-Peer Netw. Appl., 1–16 (2016)

Flexible Cell Selection in Cellular Networks

Dror Rawitz[1] and Ariella Voloshin[2]([✉])

[1] Faculty of Engineering, Bar-Ilan University, 5290002 Ramat Gan, Israel
dror.rawitz@biu.ac.il
[2] Department of Computer Science, Technion, 3200003 Haifa, Israel
variella@cs.technion.ac.il

Abstract. We introduce the problem of FLEXIBLE SCHEDULING ON RELATED MACHINES WITH ASSIGNMENT RESTRICTIONS (FSRM). In this problem the input consists of a set of machines and a set of jobs. Each machine has a finite capacity, and each job has a resource requirement interval, a profit per allocated unit of resource, and a set of machines that can potentially supply the requirement. A feasible solution is an allocation of machine resources to jobs such that: (i) a machine resource can be allocated to a job only if it is a potential supplier of this job, (ii) the amount of machine resources allocated by a machine is bounded by its capacity, and (iii) the amount of resources that are allocated to a job is either in its requirement interval or zero. Notice that a job can be serviced by multiple machines. The goal is to find a feasible allocation that maximizes the overall profit. We focus on r-FSRM in which the required resource of a job is at most an r-fraction of (or r times) the capacity of each potential machine. FSRM is motivated by resource allocation problems arising in cellular networks and in cloud computing. Specifically, FSRM models the problem of assigning clients to base stations in 4G cellular networks. We present a 2-approximation algorithm for 1-FSRM and a $\frac{1}{1-r}$-approximation algorithm for r-FSRM, for any $r \in (0,1)$. Both are based on the local ratio technique and on maximum flow computations. We also present an LP-rounding 2-approximation algorithm for a flexible version of the GENERALIZED ASSIGNMENT PROBLEM that also applies to 1-FSRM. Finally, we give an $\Omega(\frac{r}{\log r})$ lower bound on the approximation ratio for r-FSRM (assuming P $\neq$ NP).

1 Introduction

We consider the problem of FLEXIBLE SCHEDULING ON RELATED MACHINES WITH ASSIGNMENT RESTRICTIONS (abbreviated, FSRM). An FSRM instance $\mathcal{I}$ consists of a pair (I, J), where I is a set of machines and J is a set of jobs. Each machine $i \in I$ has a positive capacity denoted by $c(i)$. Each job $j \in J$ is associated with a resource demand range $[a(j), b(j)]$, where $a(j) \geq 0$, and a set of machines $N(j) \subseteq I$ that can potentially supply resources to j. A feasible solution is an allocation of machine resources to jobs, i.e., a function $x : I \times J \to \mathbb{R}^+$

Research supported in part by Network Programming (Neptune) Consortium, Israel.
D. Rawitz—Partially supported by the Israel Science Foundation (grant no. 497/14).

M. Chrobak et al. (Eds.): ALGOSENSORS 2016, LNCS 10050, pp. 112–128, 2017.
DOI: 10.1007/978-3-319-53058-1_8

such that: (i) the resources of a machine i can be allocated to a job j, only if i is a potential supplier of this job, namely $x(i,j) > 0$ only if $i \in N(j)$; (ii) the amount of machine resources allocated by a machine i is bounded by its capacity, i.e., $\sum_j x(i,j) \leq c(i)$; and (iii) the amount of resources that are allocated to a job is either in its range or zero, namely $\sum_i x(i,j) \in [a(j), b(j)] \cup \{0\}$. Each job $j \in J$ is also associated with a non-negative profit per unit of resource $w(j)$, and the goal in FSRM is to find a feasible solution that maximizes the total profit, namely that maximizes $\sum_j w(j) \sum_i x(i,j)$. The special case of FSRM in which $a = b$ is called SCHEDULING ON RELATED MACHINES WITH ASSIGNMENT RESTRICTIONS (SRM).

Note that the term *related machines* is usually used to describe a setting in which machines have different speeds, and if a job is assigned to a machine, then the load that is placed on the machine is the job demand divided by its speed. In this paper we use different capacities instead of using different speeds.

As mentioned above, in FSRM a job can be serviced by multiple machines. Such a solution is called a *cover-by-many*. A feasible solution in which each job is served by at most one machine is called a *cover-by-one*. Given a constant $r > 0$, an FSRM instance is said to be *r-restricted* if $b(j) \leq r \cdot c(i)$, for every $j \in J$ and $i \in N(j)$, i.e., no job demands more than an r-fraction of (or r times) the capacity of a machine in $N(j)$. Given $r > 0$, The variant of FSRM (SRM) where only r-restricted instances are considered is denoted by r-FSRM (r-SRM).

SRM and its variants naturally arise in resource allocation applications where clients need service from servers. Amzallag et al. [2] considered the CELL SELECTION problem which is the problem of assigning clients to base stations in 4G cellular networks where services offered by the providers (such as video streaming and web browsing) require high bit-rates, and client bandwidth requirements are non-uniform. They modeled cell selection using SRM, thus taking into account both base stations diversity, by using non-uniform capacities for each base station, as well as clients diversity, by using different demands, different profits, and different potential set of base stations.

Flexible resource allocation models the common setting in which a client's demand for resources may be partly satisfied as long as a minimal amount of resources is provided. The resource in question may be processing power, memory, storage, or communication bandwidth. For example, in 4G cellular networks, a client may be willing to pay for a certain level of video streaming in HD format, but will also settle for a lower level of quality of service (QoS) thus video streaming in SD format provided that it is not too low and that the price is reduced as well. Another application is resource allocation in the cloud. In this case machines represent cloud servers (e.g., storage servers) and jobs represent possible clients that require services from servers. Each client has a lower and upper bound on the amount of service it requires, a profit per allocated resource unit, and a list of potential servers (e.g., due to latency issues).

Related work. Both SRM and FSRM are NP-hard, since even the case where there is only one machine contains KNAPSACK as a special case. Since KNAPSACK remains NP-hard even if the size of each item is at most an r-fraction of the knapsack size, this hardness result applies to r-SRM and r-FSRM, for any $r > 0$. (This was explicitly shown in [2].)

In MULTIPLE KNAPSACK WITH ASSIGNMENT RESTRICTIONS (MKAR) the input consists of a set of bins and a set of items. Each bin has a capacity, and each item j has a size, a profit, and a subset of bins in which it can be placed. A feasible solution is an assignment of items to bins such that each item is assigned to one of the bins in its subset and the total size of items that are assigned to a bin is at most its capacity. The special case, where the size and profit of an item are the same, is considered in [7]. They presented an LP-rounding 2-approximation algorithm, a $(2+\varepsilon)$-approximation algorithm that employs an FPTAS for solving a single knapsack problem, and a greedy 3-approximation algorithm. Notice that MKAR can be seen as a job scheduling problem on related machines. However, it is not identical to SRM, since in the former a solution must be a *cover by one*.

Fleischer et al. [9] studied the SEPARABLE ASSIGNMENT PROBLEM (SAP) in which the input consists of a set of bins and a set of items, and profits p_{ij} for assigning item j to bin i. There is also a separate packing constraint for each bin, i.e., a collection $\mathcal{I}_i$ of subsets of items that fit in bin i. The goal is to maximize the total profit. Given an α-approximation algorithm for the single bin version, they presented an LP-rounding based $\frac{\alpha e}{e-1}$-approximation algorithm and a local search $(\frac{\alpha+1}{\alpha} + \varepsilon)$-approximation algorithm, for any $\varepsilon > 0$. Thus, if the single bin version admits a PTAS, then the ratios are $\frac{e}{e-1} + \varepsilon$ and $2+\varepsilon$. Moreover, the ratio drops to $\frac{e}{e-1}$, if the single bin version admits a FPTAS. In the GENERALIZED ASSIGNMENT PROBLEM (GAP) the input consists of a set of bins and a set of items. Each bin has a capacity, and each item j has a size and a profit for each bin i. A feasible solution is an assignment of items to bins such that the total size of items that are assigned to a bin is at most its capacity. GAP is a special case of SAP where the single bin version (KNAPSACK) admits an FPTAS, and thus it has an $\frac{e}{e-1}$-approximation algorithm. The best known result for GAP is an LP-rounding based $(\frac{e}{e-1} - \varepsilon)$-approximation algorithm for some constant $\varepsilon > 0$ due to Feige and Vondrák [8]. MKAR is a special case of GAP, and MULTIPLE KNAPSACK is a special case of MKAR in which each item can be placed in all bins. Chekuri and Khanna [6] gave a PTAS for MULTIPLE KNAPSACK and showed that GAP is APX-hard. In addition they observed that an LP-rounding 2-approximation algorithm for the minimization version of GAP by Shmoys and Tardos [17] implies a 2-approximation algorithm for GAP. GAP can be seen as a job scheduling problem on unrelated machines. However, as opposed to SRM a feasible solution in GAP (and SAP) must be integral, namely a *cover by one*. Hence, SRM is not a special case of GAP (or SAP). Having said that, the above mentioned 2-approximation algorithm for GAP applies to 1-SRM, since the algorithm computes a solution which is 2-approximate with respect to an optimal fractional solution to an LP-relaxation.

Amzallag et al. [2] showed that SRM cannot be approximated to within a ratio better than $|J|^{1-\varepsilon}$, for any $\varepsilon > 0$, unless NP = ZPP. They also showed that r-SRM is NP-hard, for any $r > 0$. These hardness results apply to FSRM and r-FSRM. On the positive side, they presented two algorithms for r-SRM, where $r \in (0,1)$: a $\frac{2-r}{1-r}$-approximation algorithm that computes *covers by one* and a $\frac{1}{1-r}$-approximation algorithm that computes *covers by many*. Both algorithms add jobs in a greedy manner and are analyzed using local ratio. The latter also uses MAXIMUM FLOW computations in the process of adding a job. Following [2], Patt-Shamir et al. [13] presented a distributed $(\frac{2-r}{1-r} + \varepsilon)$-approximation algorithm for SRM that runs in polylogarithmic time and computes *covers-by-one*. Gurewitz et al. [10] studied an extension of SRM, in which every job $j \in J$ and machine $i \in I$ are associated with a service rate $R(i,j)$. Under the assumption that there exists some $\delta > 0$ such that if $R(i,j) > 0$ then $R(i,j) > \delta$, they presented a $\frac{1}{(1-r)\delta}$-approximation algorithm for $r \in (0,1)$. The algorithm adopt the approach taken in [2]. However, as opposed to [2], it uses solutions to linear programs in the process of adding a job. Halldórsson et al. [11] presented centralized and distributed approximation algorithms for an extension of SRM, where a client is satisfied if it receives service from k servers.

Various flexible resource allocation problems were considered in [12,14–16]. In particular, Shachnai et al. [15] used the *local ratio technique* (see Sect. 2). to design approximation algorithms for the flexible versions of the BANDWIDTH ALLOCATION and the STORAGE ALLOCATION problems. Their use of local ratio is non-standard in the sense that the computed solution is both discrete and continuous, that is a job is either rejected or it is accepted with an assignment in its request interval.

Our results. In Sect. 3 we notice that the LP-rounding 2-approximation algorithm from [6] works for the flexible version of GAP (see formal definition in Sect. 3). This result also applies to 1-FSRM.

We present a polynomial-time local ratio $\frac{1}{1-r}$-approximation algorithm for r-FSRM, for any $r \in (0,1)$, in Sect. 5. We do so by adopting the local ratio approach taken in [15] to flexible resource allocation. We note that the algorithm works even under the weaker condition where $a(j) \leq r \cdot c(i)$, for every $i \in N(j)$. The algorithm extends the $\frac{1}{1-r}$-approximation algorithm for r-SRM from [2]. The main difference between the algorithm from [2] and ours is that in the flexible setting when it is decided to service a job j, it still remains to determine what would be the level of service j will receive. Hence, as opposed to [2], we need to use an algorithm for MAXIMUM FLOW with lower bounds. The connection between FSRM and MAXIMUM FLOW and a description of our usage of a MAXIMUM FLOW algorithm are given in Sect. 4.

We also present a polynomial-time local ratio 2-approximation algorithm for 1-FSRM in Sect. 6. This algorithm is based on a resource augmentation algorithm for 1-FSRM, which computes a solution pair (i.e., two feasible solutions) that together may exceed the available resource by a factor of 2 and whose profit is at least the optimum with the original available resources. More specifically, this solution pair consists of a *cover-by-many* and a *cover-by-one*, one of which is 2-approximate. Roughly speaking, in our first algorithm we lose a factor of $\frac{1}{1-r}$, since we may have machines whose capacity is only $1 - r$ utilized. Our second algorithm bypasses this issue by allowing solutions to over-utilize machines. The over-utilization is done such that if the capacity of a machine is violated, then it is violated by a unique job. These violating jobs constitute the cover-by-one. The algorithm is also based on computations of MAXIMUM FLOW with lower bounds, but in this case we utilize two flow networks, one for each solution. We note that, as opposed to the first 2-approximation algorithm, our second 2-approximation algorithm relies on combinatorial methods, namely the local ratio technique and MAXIMUM FLOW computations.

Our algorithms achieve the same approximation ratios as the best known for the non-flexible variants. Note that by combining our algorithms we achieve a ratio of $\min\{\frac{1}{1-r}, 2\}$ for r-FSRM, for any $r \in (0, 1]$.

Finally, we provide an $\Omega(\frac{r}{\log r})$ lower bound on the approximation ratio of r-FSRM assuming that $P \neq NP$ which was omitted for lack of space.

2 Preliminaries

Definitions and notation. Given an undirected graph $G = (V, E)$, let $E(v)$ be the set of edges incident on a vertex $v \in V$, and let $E(U) = \bigcup_{v \in U} E(v)$, for any subset $U \subseteq V$. Denote $E(U, W) = \{(u, w) : (u, w) \in E \cap (U \times W)\}$, for any $U, W \subseteq V$. For a directed graph G, let $E^-(v)$ and $E^+(v)$ be the sets of incoming arcs and outgoing arcs of a vertex v, respectively. Similarly, for any $U \subseteq V$, let $E^-(U) = \bigcup_{v \in U} E^-(v)$, let $E^+(U) = \bigcup_{v \in U} E^+(v)$, and let $E(U, W) = E^+(U) \cap E^-(W)$, for any $U, W \subseteq V$.

We consider the following model. An FSRM instance $\mathcal{I}$ is viewed as a bipartite graph $G = (I, J, E)$, where the machines are modeled by the set of vertices I, the jobs are modeled by the set of vertices J, and an edge $(i, j) \in E$ represents the fact that job j can be serviced by machine i, i.e., $(i, j) \in E$ if and only if $i \in N(j)$, where $N(j)$ is the set of machines that can potentially service j.

Given an FSRM instance, a solution $x : E \mapsto \mathbb{R}^+$ is called a *resource allocation*. Define $x(i) \triangleq \sum_{j:(i,j) \in E} x(i, j)$, for a machine $i \in I$, and $x(j) \triangleq \sum_{i:(i,j) \in E} x(i, j)$, for a job $j \in J$. Recall that a resource allocation x is feasible if $x(j) \in \{0\} \cup [a(j), b(j)]$, for every $j \in J$, and $x(i) \in [0, c(i)]$, for every $i \in I$. The total profit of a feasible resource allocation x of instance is defined as $w(x) \triangleq \sum_{j \in J} w(j)x(j)$, and thus the goal in FSRM is to find a feasible resource allocation x that maximizes $w(x)$. Given a feasible resource allocation x, the set of jobs that are serviced by x is denoted by J_x, namely $J_x = \{j \in J : x(j) \geq a(j)\}$. A *resource assignment* to jobs is a function $d : J \to \mathbb{R}^+$. Such an assignment is feasible if there is a feasible resource allocation x such that $d(j) = x(j)$, for every $j \in J$. The set J_d is defined similarly to J_x, i.e., $J_d = \{j \in J : d(j) \geq a(j)\}$.

Local ratio. The local-ratio technique [3–5] is based on the Local-Ratio Lemma, which applies to maximization problems of the following type. (See, e.g. [4] for the minimization case.) The input is a non-negative profit vector $w \in \mathbb{R}^n$ and a set $\mathcal{F}$ of feasibility constraints. The problem is to find a vector $x \in (\mathbb{R}^+)^n$ that maximizes the inner product $w \cdot x$ subject to the constraints $\mathcal{F}$.

Lemma 1 (Local Ratio [3]). *Let $\mathcal{F}$ be a set of constraints and let w, w_1, and w_2 be profit vectors such that $w = w_1 + w_2$. Also, let $\rho \geq 1$. Then, if x*

is ρ-approximate with respect to $(\mathcal{F}, w_1)$ and with respect to $(\mathcal{F}, w_2)$, then it is ρ-approximate solution with respect to $(\mathcal{F}, w)$.

3   2-Approximation Algorithm via LP-rounding

We present a 2-approximation algorithm for 1-FSRM that is based on LP-rounding. More specifically, we show that the LP-rounding 2-approximation algorithm for GAP from [6] also works for the flexible version of GAP. This result also applies to 1-FSRM.

Recall that a GAP instance consists of a set of bins and a set of items. Each bin (machine) i has a capacity $c(i)$, and each item (job) j has a size $s(i, j)$ and a profit $p(i, j)$, for each bin i. A feasible solution is an assignment of items to bins such that the total size of items that are assigned to a bin is at most its capacity. The goal is to find a maximum profit feasible solution. Chekuri and Khanna [6] (see also [17]) presented a 2-approximation algorithm for GAP that is based on LP-rounding of an LP-relaxation of the following formulation of GAP:

$$
\begin{aligned}
\max \quad & \textstyle\sum_{i,j} p(i,j) z(i,j) \\
\text{s.t.} \quad & \textstyle\sum_j s(i,j) z(i,j) \le c(i) && \forall i \\
& \textstyle\sum_i z(i,j) \le 1 && \forall j \\
& z(i,j) \in \{0,1\} && \forall i, j
\end{aligned}
\tag{GAP}
$$

FLEXIBLE GAP (FGAP) is the extension of GAP, in which each item j is associated with an interval $[a(i,j), b(i,j)]$ for its size in bin i. A formulation of FGAP is obtained by replacing $s(i,j)$ with $b(i,j)$ and by replacing the second and third sets of constraints as follows:

$$
\begin{aligned}
\max \quad & \textstyle\sum_{i,j} p(i,j) z(i,j) \\
\text{s.t.} \quad & \textstyle\sum_j b(i,j) z(i,j) \le c(i) && \forall i \\
& z(i,j) \in \{0\} \cup [\tfrac{a(i,j)}{b(i,j)}, 1] && \forall i, j \\
& z(i,j) > 0 \text{ for at most one } i && \forall j
\end{aligned}
\tag{FGAP}
$$

Observe that the LP-relaxation of (GAP), in which the integrality constraints are replaced by $z(i,j) \ge 0$ for every i and j, is also a relaxation of (FGAP). Since the algorithm from [6] computes a cover-by-one, the solution is feasible with respect to (FGAP). This leads to the following result.

Theorem 1. *There is an LP-rounding 2-approximation algorithm for FGAP.*

The LP-relaxation of (FGAP) is also a relaxation of 1-FSRM: a resource allocation x is obtained by assigning: $x(i,j) = z(i,j) b(i,j) = z(i,j) b(j)$, for every i and j. In this case $p(i,j) = w(i,j) b(i,j)$.

Corollary 1. *There is an LP-rounding 2-approximation algorithm for 1-FSRM that computes a cover-by-one x where $x(i,j) \in \{0, b(j)\}$, for every j.*

4   FSRM and Maximum Flow

In this section we represent resource allocations using flow and design algorithms to compute resource allocations using a MAXIMUM FLOW algorithm. The representation and the algorithms are the building blocks of our approximation algorithms. Throughout the section we use an algorithm, denoted by MAXFLOWLB, that computes a maximum flow in a network with both upper and lower bounds on the edges (see, e.g., [1, Sect. 6.7]). We also assume that the computed flow is integral, given integral capacities.

4.1   Representing Resource Allocations Using Flow

Given an FSRM instance $\mathcal{I}$, consider the following standard flow network. Starting with the graph $G = (I, J, E)$, add a source s, a target t, and directed edges from s to the machines in I and from the jobs in J to t. In addition, direct the edges in E from I to J. More formally, construct a directed graph $H = (V, E')$, where $V = \{s, t\} \cup I \cup J$, and $E' = \{(i, j) : i \in I, j \in J, (i, j) \in E\} \cup \{(s, i) : i \in I\} \cup \{(j, t) : j \in J\}$.

We also define lower and upper bounds, ℓ and u, resp., on the edges as follow (ℓ is defined for completeness):

$$\ell = 0 \qquad u(v, v') = \begin{cases} c(v') & v = s, v' \in I, \\ \infty & (v, v') \in E, \\ b(v) & v \in J, v' = t. \end{cases}$$

Given an FSRM instance $\mathcal{I}$, observe that a feasible resource allocation x induces the following feasible flow on the above flow network:

$$f_x(v, v') = \begin{cases} x(v') & v = s, v' \in I, \\ x(v, v') & v \in I, v' \in J, \\ x(v) & v \in J, v' = t. \end{cases}$$

For the other direction, a feasible flow f induces a feasible resource allocation x, if $f(j, t) \notin (0, a(j))$, for every job $j \in J$. Let x_f be the resource allocation that is induced by f, namely $x_f(i, j) = f(i, j)$, for every $(i, j) \in E$. Given a flow f let $|f|$ denote the amount of flow carried by f from s to t, namely let $|f| = \sum_i f(s, i)$. Given a solution x, $|f_x|$ stands for the amount of resources used by x. A maximum flow in this network is denoted by f^*. Note that f^* may not correspond to a feasible resource allocation function x.

Finally, given two subsets $I' \subseteq I$ and $J' \subseteq J$, the network that is induced by I', J', and $E(I', J')$ is denoted by $H(I', J')$.

4.2   Extending Solutions

In this section we define the notion of *maximal* resource allocation and show how to extend such an allocation.

Algorithm 1. Extend $((I, J, E), a, b, c, x, k)$

1: $f \leftarrow \text{MaxFlowLB}(H, \ell^0, u^0)$ ▷ $f = \emptyset$ if no flow exists
2: **if** $f \neq \emptyset$ **then return** $f(k, t)$
3: **else return** 0

Definition 1. *Given an FSRM instance* $\mathcal{I}$, *a feasible resource allocation* x *is called* maximal *if there is no feasible resource allocation* y *such that* $y(j) \geq x(j)$, *for every job* $j \in J$, *and* $y(j') > x(j')$, *for some job* $j' \in J$. *A resource assignment* $d : J \to \mathbb{R}^+$ *is called* maximal *if there exists a maximal resource allocation* x *such that* $d(j) = x(j)$.

We present an algorithm, called **Extend** (Algorithm 1), whose input is an FSRM instance (i.e., (I, J, E), minimum and maximum resource demand functions a and b, and a capacity function c), a job $k \in J$, and a resource allocation x, which is maximal with respect to $(I, J\backslash\{k\}, E\backslash E(k))$. The algorithm tries to compute a maximal resource allocation with respect to (I, J, E) that extends x using k. In other words, it checks whether k can be feasibly added to the solution, such that the resource allocation of each job $j \in J\backslash\{k\}$ does not change, and if it can, it returns the maximum resource assignment it can receive. Algorithm **Extend** uses MaxFlowLB with the derived flow network H and the following bounds on the edges:

$$\ell^0(v, v') = \begin{cases} x(v) & v \in J\backslash\{k\}, v' = t, \\ a(k) & v = k, v' = t, \\ 0 & \text{otherwise.} \end{cases} \qquad u^0 = u$$

It is not hard to verify that the running time of **Extend** is dominated by the running time of the Maximum Flow algorithm.

Lemma 2. *Let* $((I, J, E), a, b, c, x, k)$ *be the input of **Extend**, let* $d(k)$ *be the value returned by **Extend** and let* $d(j) = x(j)$, *for every* $j \in J\backslash\{k\}$. *If* x *is maximal with respect to* $(I, J\backslash\{k\}, E\backslash E(k))$, *then* d *is maximal with respect to* (I, J, E).

Proof. Assume towards contradiction that d is not maximal, and there exists a feasible resource assignment d' such that $d'(j) \geq d(j)$, for every job $j \in J$, and $d'(j') > d(j')$, for some job $j' \in J$. Observe that $f(j) = x(j)$, for $j \neq k$, since $\ell^0(j, t) = x(j)$ and x is maximal. Hence $d(j) = d'(j)$, for $j \neq k$. It follows that MaxFlowLB computes the flow that maximizes the flow on the edge (k, t), as long as it is in the range $[a(k), b(k)]$. Hence, $d'(k) \leq d(k)$. A contradiction. $\square$

4.3 Creating Flow-Maximal Solutions

In this section we define the notion of a *flow-maximal* solution and present an algorithm that computes such a solution.

Algorithm 2. Assign $((I, J, E), b, c, d, M)$

1: $f \leftarrow$ MAXFLOWLB(H, ℓ^1, u^1)
2: $g \leftarrow$ MAXFLOWLB(H, ℓ^2, u^2) ▷ $g = \emptyset$ if no flow exists
3: **if** $g \neq \emptyset$ **then return** x_g
4: **else return** FAIL

Definition 2. *Given an* FSRM *instance* $\mathcal{I}$, *a resource allocation* x *is called flow-maximal with respect to a subset of machines* $M \subseteq I$, *if (i)* x *is maximal; and (ii)* $f_x(s, i) = f^*(s, i)$, *for every* $i \in M$, *where* f^* *is a maximum flow in the flow network* $H(M, J_x)$ *with upper bounds* u.

Algorithm **Assign** (Algorithm 2) computes a flow-maximal solution with respect to a given subset of machines. The input consists of an FSRM instance (i.e., (I, J, E), a maximum resource demand function b, a capacity function c), a subset of machines $M \subseteq I$, and a maximal resource assignment d. The goal is to compute a flow-maximal resource allocation y with respect to M, such that $y(j) = d(j)$, for every $j \in J$, if such an allocation exists. **Assign** has two phases. In the first it tries to get as much service as possible from machines in M. To that end it computes a maximum flow f for the following edge bounds:

$$\ell^1 = 0 \qquad u^1(v, v') = \begin{cases} 0 & v = s, v' \notin M, \\ 0 & v \in J, v' = t, d(v) = 0 \\ u(v, v') & \text{otherwise.} \end{cases}$$

In the second phase **Assign** uses Algorithm MAXFLOWLB with the goal of maintaining the resources that f supplies using machines from M. **Assign** computes a maximum flow for the following bounds:

$$\ell^2(v, v') = \begin{cases} f(v, v') & v = s, v' \in M \\ d(v) & v \in J, v' = t, \\ 0 & \text{otherwise.} \end{cases} \qquad u^2 = u.$$

As **Extend**, the running time of **Assign** is dominated by the running time of the MAXIMUM FLOW algorithm.

We show that if **Assign** receives a maximal resource assignment d as input, then it computes a flow-maximal solution x_g with respect to M. We first show that starting with f is not a bad move.

Lemma 3. *Let* $((I, J, E), b, c, d, M)$ *be the input of* **Assign**, *where* d *is a maximal resource assignment. Then there exists a maximal resource allocation* x *such that* $x(j) = d(j)$, *for every* $j \in J$, *and* $f_x(s, i) \geq f(s, i)$, *for every* $i \in I$.

Proof. Since d is maximal, there exists a maximal resource allocation x such that $x(j) = d(j)$, for every job $j \in J$. Let x be such a maximal resource allocation that maximizes: $\Delta(x) \triangleq \sum_{i \in M, j \in J_x} \min \{f(i, j), f_x(i, j)\}$. If $f_x(s, i) \geq f(s, i)$, for every $i \in M$, then we are done. Otherwise, there exists a machine $q \in M$ such that $f(s,q) > f_x(s,q)$, and it follows that there exists a job j such that $f(q,j) > f_x(q,j)$.

If $d(j) < b(j)$, one may increase $f_x(s,q)$, $f_x(q,j)$, and $f_x(j,t)$ by the value $\min\{f(s,q) - f_x(s,q), b(j) - d(j)\}$ in contradiction to the maximality of x and d. If $d(j) = b(j)$, then there must exist a machine q' such that $f_x(q',j) > f(q',j)$. Construct a new solution x' by letting q service j instead of q' in the amount $\min\{f(s,q) - f_x(s,q), f(q,j) - f_x(q,j), f_x(q',j) - f(q',j)\}$, where $f(s,q) - f_x(s,q)$ is the available resource of q, $f(q,j) - f_x(q,j)$ is the service gap of j, and $f_x(q',j) - f(q',j)$ is the amount of resource q' gives j. Hence $x'(j') = d(j')$, for every j', and $\Delta(x') > \Delta(x)$. A contradiction to the maximality of $\Delta(x)$.   □

Next we show that the computed solution is flow-maximal.

Lemma 4. *Let $((I, J, E), b, c, d, M)$ be the input of **Assign**, where d is a maximal resource assignment. Then $x_g(j) = d(j)$, for every $j \in J$, and $g(s,i) = f(s,i)$, for every $i \in M$.*

Proof. By Lemma 3 there exists a maximal resource allocation x such that $x(j) = d(j)$, for every $j \in J$, and $f_x(s,i) \geq f(s,i)$, for every $i \in I$. Hence Line 2 succeeds in computing a flow g with the bounds ℓ^2 and u^2. Moreover, d is maximal and the construction of ℓ^2 and u^2 implies that $x_g(j) = d(j)$, for every $j \in J$, and that $g(s,i) \geq f(s,i)$, for every $i \in I$. Since f is a maximum flow in $H(M, J_{x_g})$ it must be that $g(s,i) = f(s,i)$, for every $i \in M$.   □

5   A $\frac{1}{1-r}$-Approximation Algorithm for r-FSRM

In this section we present a polynomial-time $\frac{1}{1-r}$-approximation algorithm for r-FSRM, where $r \in (0,1)$. Our algorithm is based on the local ratio technique, and it uses Algorithms **Extend** and **Assign** as subroutines.

We first introduce notations that we use in this section. Given an FSRM instance $\mathcal{I}$, and a subset $K \subseteq J$, let $N(K) = \bigcup_{j \in K} N(j)$ be the set containing machines that can potentially serve jobs in K, and let $\bar{N}(K) = I \backslash N(K)$ be the set of machines that cannot serve jobs in K. We also denote $\bar{\bar{N}}(K) = \bar{N}(J \backslash K)$, where $\bar{\bar{N}}(K)$ contains the machines that can only serve jobs in K. Figure 1 depicts these notations.

Algorithm **FSRM-LR** (Algorithm 3) receives an FSRM instance. It is recursive and works as follows. If there are no jobs, then it returns an empty resource allocation. Otherwise, it chooses a job k with minimum profit. It constructs a new profit function w_1, which assigns a profit of $w(k)$ to all jobs and solves the problem recursively on $(I, J \backslash \{k\}, E \backslash E(k))$ and $w_2 = w - w_1$. Then, it calls Algorithm **Extend** to check whether k can be added to the solution that was computed recursively while maintaining feasibility. Following this resource assignment, it calls Algorithm **Assign** that constructs a flow-maximal resource allocation with respect to the machines that can only serve the assigned jobs

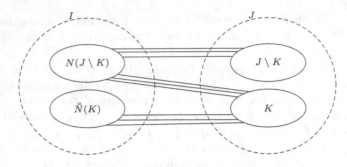

Fig. 1. Definition of $\bar{\bar{N}}(K)$.

Algorithm 3. FSRM-LR $((I, J, E), w)$

1: **if** $J = \emptyset$ **then return** an empty resource allocation
2: $k \leftarrow \text{argmin}_{j \in J}\, w(j)$
3: Define: $w_1(j) = w(k)$, for every $j \in J$, and $w_2 = w - w_1$
4: $x \leftarrow$ **FSRM-LR**$((I, J\backslash\{k\}, E\backslash E(k)), w_2)$
5: $d(j) \leftarrow x(j)$, for every $j \in J\backslash\{k\}$
6: $d(k) \leftarrow$ **Extend**$((I, J, E), a, b, c, x, k)$
7: $y \leftarrow$ **Assign**$((I, J, E), b, c, d, \bar{N}(J_d))$
8: **return** y

---

and the new resource assignment. Observe that even though the algorithm is recursive and uses profit functions, it actually scans jobs in a non-decreasing order of profit per unit (i.e., Line 2 chooses a minimum profit job among the remaining jobs). Still, we use recursion and these functions since they simplify the local ratio analysis.

We start the analysis of **FSRM-LR** by showing that uniform profit functions (i.e., all jobs have the same profit) are useful from a local ratio point of view.

Lemma 5. *Given an r-FSRM instance $\mathcal{I}$, if let y is flow-maximal with respect to $\bar{\bar{N}}(J_y)$, then y is $\frac{1}{1-r}$-approximate with respect to a uniform profit function.*

Proof. Since y is flow-maximal with respect to $\bar{\bar{N}}(J_y)$, $\sum_{i \in \bar{\bar{N}}(J_y)} f_y(s, i) = |f^*|$, where f^* is a maximum flow in the flow network $H(\bar{\bar{N}}(J_y), J_y)$. Let z be any feasible resource allocation of (I, J, E). Hence,

$$|f_z| = \sum_{i \in \bar{N}(J_y)} f_z(s, i) + \sum_{i \in N(J\backslash J_y)} f_z(s, i) \leq |f^*| + c(N(J\backslash J_y)) = \sum_{i \in \bar{N}(J_y)} f_y(s, i) + c(N(J\backslash J_y))$$

Also notice that since $a(j) \leq r \cdot c(i)$, for every $(i, j) \in E$, we have that $f_y(s, i) = y(i) \geq (1 - r)c(i)$, for every $i \in N(J\backslash J_y)$. Otherwise, there exists a job $j \in J\backslash J_y$ that can be assigned to a machine $i \in N(J\backslash J_y)$ in contradiction to the maximality of y. (This claim is vacuously correct if $J\backslash J_y = \emptyset$.) It follows that

$$|f_z| \leq \sum_{i \in \bar{N}(J_y)} f_y(s,i) + \frac{1}{1-r} \sum_{i \in N(J \setminus J_y)} f_y(s,i) \leq \frac{1}{1-r} \sum_{i \in I} f_y(s,i) = \frac{1}{1-r} |f_y| .$$

The lemma follows due to the uniform profits.   □

It remains to prove that **FSRM-LR** is a $\frac{1}{1-r}$-approximation algorithm.

Theorem 2. *Algorithm* **FSRM-LR** *is a polynomial-time $\frac{1}{1-r}$-approximation algorithm for r-FSRM, for $r \in (0,1)$.*

Proof. The proof is by induction on the number of recursive calls, where in addition we show that the returned solution y is flow-maximal with respect to $\bar{N}(J_y)$. At the recursion base, the solution returned is optimal, $\frac{1}{1-r}$-approximate, and flow-maximal, since $J = \emptyset$. For the inductive step, x is flow-maximal with respect to $\bar{N}(J_x)$ by the induction hypothesis. By Lemmas 2 and 4 y is also flow-maximal with respect to $\bar{N}(J_y)$. Next, we show that the computed solution y is $\frac{1}{1-r}$-approximate with respect to J and both w_1 and w_2. x is $\frac{1}{1-r}$-approximate with respect to $J \setminus \{k\}$ and w_2 due to the inductive hypothesis. $w_2(k) = 0$, and $y(j) = x(j)$ for every $j \in J \setminus \{k\}$, due to Lemma 4. Hence y is $\frac{1}{1-r}$-approximate with respect to J and w_2. Since y is flow-maximal with respect to $\bar{N}(J_y)$ and w_1 is a uniform profit function, y is $\frac{1}{1-r}$-approximate with respect to J and w_1 by Lemma 5. By the Local Ratio Lemma (Lemma 1) we have that y is $\frac{1}{1-r}$-approximate with respect to J and w.

Finally, the running time of **FSRM-LR** is polynomial, since the number of recursive call is $|J|$, and the running time of each such call is polynomial, as **Extend** and **Assign** terminate in polynomial time.   □

We note that our analysis is tight even in the case of SRM. Let $t \in \mathbb{N}$ and consider an SRM instance that consists of a single machine, with capacity 1, and $2t-1$ jobs, where $a(j) = b(j) = \frac{1}{t}$ and $w(j) = 1$, for $j \leq t$, and $a(j) = b(j) = \frac{1+\varepsilon}{t}$ and $w(j) = 1$, for $j > t$, where $\varepsilon > 0$ is a small number. The algorithm may compute a solution containing jobs $t+1, \ldots, 2t-1$ whose profit is $\frac{(t-1)(1+\varepsilon)}{t}$, while the profit of an optimal solution is 1. The ratio is $\frac{1}{(1+\varepsilon)-r}$, where $r = \frac{1+\varepsilon}{t}$.

6   A 2-Approximation Algorithm for 1-FSRM

We present a 2-approximation algorithm for 1-FSRM. It is based on a resource augmentation algorithm for 1-FSRM which maintains two feasible resource allocations, each using its own set of machines. We show that the combined profit of these solutions is at least the optimum profit. A 2-approximation is obtained by choosing the one with the higher profit.

Before presenting the algorithm we define the notion of *auxiliary instance*. Given an FSRM instance $\mathcal{I}$, its auxiliary instance contains the same machine set I and job set J. However, all machine capacities are unit, and all jobs lower and upper bounds are unit. That is, we define $c^A(i) = 1$, for every $i \in I$, and $a^A(j) = b^A(j) = 1$, for every $j \in J$. The flow upper bound u with respect to c^A

Algorithm 4. FSRM-RA $((I, J, E), w)$

1: **if** $J = \emptyset$ **then return** two empty resource allocations
2: $k \leftarrow \operatorname{argmin}_{j \in J} w(j)$
3: Define: $w_1(j) = w(k)$, for every $j \in J$, and $w_2 = w - w_1$
4: $(x, x^A) \leftarrow$ **FSRM-RA**$((I, J\backslash\{k\}, E\backslash E(k)), w_2)$
5: $d(j) \leftarrow x(j)$ and $d^A(j) \leftarrow x^A(j)$, for every $j \in J\backslash\{k\}$
6: $d(k) \leftarrow$ **Extend**$((I, J, E), a, b, c, x, k)$
7: **if** $d(k) = 0$ **then** $d^A(k) \leftarrow$ **Extend**$((I, J, E), a^A, b^A, c^A, x^A, k)$
8: **else** $d^A(k) = 0$
9: $y \leftarrow$ **Assign**$((I, J, E), b, c, d, \bar{\bar{N}}(J_d \cup J_{d^A}^A))$
10: $y^A \leftarrow$ **Assign**$((I, J, E), b^A, c^A, d^A, \bar{\bar{N}}(J_d \cup J_{d^A}^A))$
11: **return** (y, y^A)

---

and b^A is denoted by u^A. Observe that, in this case, an integral flow corresponds to a *matching* of machines to jobs.

Algorithm **FSRM-RA** (Algorithm 4) is based on Algorithm **FSRM-LR**. However, as opposed to **FSRM-LR** it maintains two feasible solutions. The first is a *cover-by-many* similar to the one constructed by **FSRM-LR**, and the second is a *cover-by-one* that is constructed using the auxiliary instance. A job may be served by at most one of the solutions. More formally, the output is a pair (y, y^A), where y is a feasible resource allocation of the original instance, and $y^A :$ $E \mapsto \{0, 1\}$ is an integral feasible resource allocation of the auxiliary instance. y^A is called the *auxiliary solution*. We note that y and y^A are constructed such that y is flow-maximal solution with respect to $\bar{\bar{N}}(J_y \cup J_{y^A}^A)$ in $H(I, J)$ with u, while y^A is flow-maximal with respect to $\bar{\bar{N}}(J_y \cup J_{y^A}^A)$ in $H(I, J\backslash J_y)$ with u^A. Such a pair (y, y^A) of resource allocations is called a *flow-maximal pair* with respect to $\bar{\bar{N}}(J_y \cup J_{y^A}^A)$. Given y^A, a *cover-by-one* for the original instance can be obtained by $\hat{y}(i, j) \triangleq b(j) \cdot y^A(i, j)$, for every $(i, j) \in E$. Here we use the assumption that $r = 1$ and thus $b(j) \leq c(i)$, for every $(i, j) \in E$. Notice that $J_{\hat{y}} = J_{y^A}^A \triangleq \{j \in J : y^A(j) = 1\}$.

FSRM-RA is a recursive local ratio algorithm and it works as follows. If there are no jobs, then it returns two empty resource allocations. Otherwise, it chooses a job k with minimum profit. It constructs a new profit function w_1, which assigns profit of $w(k)$ to all jobs and solves the problem recursively on $(I, J\backslash\{k\}, E\backslash E(k))$ and $w_2 = w - w_1$. Let (x, x^A) be this computed recursive solution. Then, it calls Algorithm **Extend** to check whether k can be added to the solution x. If x can not be extended, it calls **Extend** to check whether k can be added to the auxiliary solution x^A. Then Algorithm **Assign** is called for each instance to reconstruct a flow-maximal allocations with respect to the machines that can only serve the assigned jobs of both solutions.

In order to prove that the computed solution is 1-approximate with respect to w_1, we show that the combined flow of y and $\hat{y}$ is at least as high as an optimal flow with respect to $H(\bar{\bar{N}}(J_y \cup J_{\hat{y}}), J_y \cup J_{\hat{y}})$ and u.

Lemma 6. *Given a 1-FSRM instance $\mathcal{I}$, let (y, y^A) be a flow-maximal pair with respect to $\bar{\bar{N}}(J_y \cup J_{y^A}^A)$. Then, $\sum_{i \in \bar{N}(J_y \cup J_{\hat{y}})}[y(i) + \hat{y}(i)] \geq |f^*|$, where f^* is a maximum flow in the network $H(\bar{\bar{N}}(J_y \cup J_{\hat{y}}), J_y \cup J_{\hat{y}})$ with upper bounds u.*

Proof. Let (S, T) be a minimum cut with respect to the flow network $H(\bar{\bar{N}}(J_y \cup J_{\hat{y}}), J_y)$ and the upper bounds u. Let (S^A, T^A) be a minimum cut that corresponds to f_{y^A} with respect to the flow network $H(\bar{\bar{N}}(J_y \cup J_{\hat{y}}), J_{\hat{y}}))$ and the auxiliary bounds u^A. We construct a new cut $(\tilde{S}, \tilde{T})$ in the flow network $H(\bar{\bar{N}}(J_y \cup J_{\hat{y}}), J_y \cup J_{\hat{y}})$ and u as follows: $\tilde{T} = T \cup T^A$ and $\tilde{S} = V \backslash \tilde{T}$, where $V = \bar{\bar{N}}(J_y \cup J_{\hat{y}}) \cup J_y \cup J_{\hat{y}} \cup \{s, t\}$. We have that $u(\tilde{S}, \tilde{T})$ is equal to

$$u(\tilde{S}, \tilde{T}) = \sum_{v \in \tilde{S}, v' \in \tilde{T}} u(v, v') = \sum_{i \in I \cap \tilde{T}} u(s, i) + \sum_{i \in I \cap \tilde{S}, j \in J \cap \tilde{T}} u(i, j) + \sum_{j \in J \cap \tilde{S}} u(j, t)$$

$$= \sum_{i \in I \cap \tilde{T}} u(s, i) + \sum_{j \in J \cap \tilde{S}} u(j, t)$$

where the second equality is because $u(i, j) = \infty$, for every $(i, j) \in E$

Consider a machine $i \in \tilde{T}$. If $i \in T$, then $f_y(s, i) = u(s, i)$ due to Lemma 4 and the max-flow min-cut Theorem. If $i \in T^A$, then $f_{y^A}(s, i) = u^A(s, i) = 1$ similarly and it follows that there is a job j such that $f_{y^A}(i, j) = 1$. Hence, $f_y(s, i) > c(i) - a(j)$ and $f_{\hat{y}}(s, i) \geq a(j)$. Therefore, in both cases $f_y(s, i) + f_{\hat{y}}(s, i) \geq u(s, i)$. Now consider a job $j \in \tilde{S}$. If $j \in J_y$, then due to Lemma 4 and min-cut max-flow Theorem $f_y(j, t) = u(j, t)$ and $f_{\hat{y}}(j, t) = 0$. On the other hand, if $j \in J_{\hat{y}}$, then $f_{\hat{y}}(j, t) = b(j) = u(j, t)$ and $f_y(j, t) = 0$. Therefore, in both cases $f_y(j, t) + f_{\hat{y}}(j, t) = u(j, t)$. Hence

$$u(\tilde{S}, \tilde{T}) \leq \sum_{i \in I \cap \tilde{T}} [f_y(s, i) + f_{\hat{y}}(s, i)] + \sum_{j \in J \cap \tilde{S}} [f_y(j, t) + f_{\hat{y}}(j, t)]$$

$$= \sum_{i \in I \cap \tilde{T}} f_y(s, i) + \sum_{j \in J \cap \tilde{S}} f_y(j, t) + \sum_{i \in I \cap \tilde{T}} f_{\hat{y}}(s, i) + \sum_{j \in J \cap \tilde{S}} f_{\hat{y}}(j, t)$$

$$= \sum_{i \in \bar{N}(J_y \cup J_{y^A}^A)} f_y(s, i) + \sum_{i \in \bar{N}(J_y \cup J_{y^A}^A)} f_{\hat{y}}(s, i) \qquad (1)$$

$$= \sum_{i \in \bar{N}(J_y \cup J_{y^A}^A)} [y(i) + \hat{y}(i)] .$$

where (1) is since the value of a flow remains the same for any s, t-cut. The lemma follows, since $|f^*| \leq u(\tilde{S}, \tilde{T})$.   □

Next we show that the computed solution pair is 1-approximate with respect to a uniform profit function.

Lemma 7. *Given a 1-FSRM instance $\mathcal{I}$, let (y, y^A) be flow-maximal pair with respect to $\bar{\bar{N}}(J_y \cup J_{y^A}^A)$. Then $(y, \hat{y})$ is 1-approximate with respect to a uniform profit function.*

Proof. Let z be any feasible resource allocation of $\mathcal{I}$. Hence,

$$|f_z| = \sum_{i \in \bar{\bar{N}}(J_y \cup J_{\hat{y}})} f_z(s,i) + \sum_{i \in N(J \setminus (J_y \cup J_{\hat{y}}))} f_z(s,i)$$

$$\leq |f^*| + c(N(J \setminus (J_y \cup J_{\hat{y}})))$$

$$\leq \sum_{i \in \bar{\bar{N}}(J_y \cup J_{\hat{y}})} [y(i) + \hat{y}(i)] + c(N(J \setminus (J_y \cup J_{\hat{y}}))) , \qquad (2)$$

where f^* is a maximum flow in the network $H(\bar{\bar{N}}(J_y \cup J_{\hat{y}}), J_y \cup J_{\hat{y}})$, and (2) is due to Lemma 6. Since $a(j) \leq c(i)$, for every $(i,j) \in E$, we have that $y(i)+\hat{y}(i) = f_y(s,i) + f_{\hat{y}}(s,i) \geq c(i)$, for every $i \in N(J \setminus (J_y \cup J_{\hat{y}}))$. Otherwise, if there exists a machine $i \in N(J \setminus (J_y \cup J_{\hat{y}}))$ for which $f_y(s,i) + f_{\hat{y}}(s,i) < c(i)$, then there is a job $j \in J \setminus (J_y \cup J_{\hat{y}})$ that can be assigned to i, in contradiction to the maximality of y and y^A. (This claim is vacuously correct if $J \setminus (J_y \cup J_{\hat{y}}) = \emptyset$.) Hence

$$|f_z| \leq \sum_{i \in \bar{\bar{N}}(J_y \cup J_{\hat{y}})} [f_y(s,i) + f_{\hat{y}}(s,i)] + \sum_{i \in N(J \setminus (J_y \cup J_{\hat{y}}))} [f_y(s,i) + f_{\hat{y}}(s,i)]$$

$$\leq \sum_{i \in I} [f_y(s,i) + f_{\hat{y}}(s,i)]$$

$$= |f_y| + |f_{\hat{y}}| .$$

The lemma follows, since the profit function is uniform. □

It remains to prove that **FSRM-RA** is a 1-approximation algorithm.

Theorem 3. FSRM-RA *is a polynomial-time algorithm that computes 1-approximations for 1-FSRM using up to twice the amount of available resources.*

Proof. The proof that the solution pair $(y, \hat{y})$ is 1-approximate is by induction on the number of recursive calls. At the base of the recursion, the return solution pair is optimal, since $J = \emptyset$. For the inductive step, we show that the computed solution pair $(y, \hat{y})$ is 1-approximate with respect to w_1 and w_2. $(x, \hat{x})$ is 1-approximate with respect to $J \setminus \{k\}$ and w_2 due to the inductive hypothesis. Since $w_2(k) = 0$, it follows that $(y, \hat{y})$ is 1-approximate with respect to J and w_2. We now show that $(y, \hat{y})$ is a 1-approximate solution with respect to w_1. By Lemma 4 y is flow-maximal with respect to $\bar{\bar{N}}(J_y \cup J_{y^A}^A)$ in $H(I, J)$ with u and y^A is flow-maximal with respect to $\bar{\bar{N}}(J_y \cup J_{y^A}^A)$ in $H(I, J \setminus J_y)$ with u^A. Therefore, (y, y^A) is flow-maximal pair with respect to $\bar{\bar{N}}(J_y \cup J_{y^A}^A)$. In addition w_1 is uniform profit function. Hence the pair $(y, \hat{y})$ is 1-approximate with respect to J and w_1 by Lemma 7. By the Local Ratio Lemma (Lemma 1) we have that $(y, \hat{y})$ is 1-approximate with respect to J and w.

Finally, the running time of **FSRM-RA** is polynomial, since the number of recursive call is $|J|$, and the running time of each such call is polynomial, as **Extend** and **Assign** terminate in polynomial time. □

Given the two feasible resource allocations computed by **FSRM-RA**, by choosing the resource allocation with the highest profit we obtain the following:

Corollary 2. *There is a polynomial-time 2-approximation algorithm for* 1-FSRM.

We show that the analysis is tight using an SRM instance with a single unit capacity machine and three jobs, where $a(1) = b(1) = 1$ and $w(1) = 1$, and $a(2) = b(2) = a(3) = b(3) = \frac{1}{2} + \varepsilon$, where $\varepsilon > 0$ is a small constant, and $w(2) = w(3) = 1$. The algorithm may pack 2 and 3, and thus output one of them, while an optimal solution consists of job 1.

References

1. Ahuja, R.K., Magnanti, T.L., Orlin, J.B.: Network Flows: Theory, Algorithms, and Applications. Prentice-Hall, Inc. (1993)
2. Amzallag, D., Bar-Yehuda, R., Raz, D., Scalosub, G.: Cell selection in 4G cellular networks. IEEE Trans. Mob. Comput. **12**(7), 1443–1455 (2013)
3. Bar-Noy, A., Bar-Yehuda, R., Freund, A., Naor, J., Schieber, B.: A unified approach to approximating resource allocation and scheduling. J. ACM **48**(5), 1069–1090 (2001)
4. Bar-Yehuda, R., Bendel, K., Freund, A., Rawitz, D.: Local ratio: a unified framework for approximation algorithms. ACM Comput. Surv. **36**(4), 422–463 (2004)
5. Bar-Yehuda, R., Even, S.: A local-ratio theorem for approximating the weighted vertex cover problem. Ann. Discrete Math. **25**, 27–46 (1985)
6. Chekuri, C., Khanna, S.: A polynomial time approximation scheme for the multiple knapsack problem. SIAM J. Comput. **35**(3), 713–728 (2005)
7. Dawande, M., Kalagnanam, J., Keskinocak, P., Salman, F.S., Ravi, R.: Approximation algorithms for the multiple knapsack problem with assignment restrictions. J. Comb. Optim. **4**(2), 171–186 (2000)
8. Feige, U., Vondrák, J.: Approximation algorithms for allocation problems: improving the factor of $1-1/e$. In: 47th FOCS, pp. 667–676 (2006)
9. Fleischer, L., Goemans, M.X., Mirrokni, V.S., Sviridenko, M.: Tight approximation algorithms for maximum general assignment problems. In: 17th SODA, pp. 611–620 (2006)
10. Gurewitz, O., Sandomirsky, Y., Scalosub, G.: Cellular multi-coverage with nonuniform rates. In: INFOCOM, pp. 1330–1338 (2014)
11. Halldórsson, M.M., Köhler, S., Rawitz, D.: Distributed approximation of k-service assignment. In: 19th OPODIS (2015)
12. Katz, D., Schieber, B., Shachnai, H.: Brief announcement: flexible resource allocation for clouds and all-optical networks. In: SPAA (2016)
13. Patt-Shamir, B., Rawitz, D., Scalosub, G.: Distributed approximation of cellular coverage. J. Parallel Distrib. Comput. **72**(3), 402–408 (2012)
14. Shachnai, H., Voloshin, A., Zaks, S.: Optimizing bandwidth allocation in flex-grid optical networks with application to scheduling. In: IPDPS, pp. 862–871 (2014)
15. Shachnai, H., Voloshin, A., Zaks, S.: Flexible bandwidth assignment with application to optical networks. In: Csuhaj-Varjú, E., Dietzfelbinger, M., Ésik, Z. (eds.) MFCS 2014. LNCS, vol. 8635, pp. 613–624. Springer, Heidelberg (2014). doi:10.1007/978-3-662-44465-8_52

16. Shalom, M., Wong, P.W.H., Zaks, S.: Profit maximization in flex-grid all-optical networks. In: Moscibroda, T., Rescigno, A.A. (eds.) SIROCCO 2013. LNCS, vol. 8179, pp. 249–260. Springer, Heidelberg (2013). doi:10.1007/978-3-319-03578-9_21
17. Shmoys, D.B., Tardos, É.: An approximation algorithm for the generalized assignment problem. Math. Program. 62, 461–474 (1993)

The Euclidean k-Supplier Problem in $I\!R^2$

Manjanna Basappa[1], Ramesh K. Jallu[1],
Gautam K. Das[1]([✉]), and Subhas C. Nandy[2]

[1] Indian Institute of Technology Guwahati, Guwahati 781039, India
gkd@iitg.ernet.in
[2] Indian Statistical Institute, Kolkata 700108, India

Abstract. In this paper, we consider k-supplier problem in $I\!R^2$. Here, two sets of points $\mathcal{P}$ and $\mathcal{Q}$ are given. The objective is to choose a subset $\mathcal{Q}_{opt} \subseteq \mathcal{Q}$ of size at most k such that congruent disks of minimum radius centered at the points in $\mathcal{Q}_{opt}$ cover all the points of $\mathcal{P}$.

We propose a fixed-parameter tractable (FPT) algorithm for the k-supplier problem that produces a 2-factor approximation result. For $|P| = n$ and $|Q| = m$, the worst case running time of the algorithm is $O(6^k(n+m)\log(mn))$, which is an exponential function of the parameter k. We also propose a heuristic algorithm based on Voronoi diagram for the k-supplier problem, and experimentally compare the result produced by this algorithm with the best known approximation algorithm available in the literature [Nagarajan, V., Schieber, B., Shachnai, H.: The Euclidean k-supplier problem, In Proc. of 16th Int. Conf. on Integ. Prog. and Comb. Optim., 290–301 (2013)]. The experimental results show that our heuristic algorithm is slower than Nagarajan et al.'s $(1 + \sqrt{3})$-approximation algorithm, but the results produced by our algorithm significantly outperforms that of Nagarajan et al.'s algorithm.

Keywords: k-supplier problem · FPT algorithm · Approximation algorithm

1 Introduction

In this paper we study a generalization of discrete k-center problem, which is known as the *k-supplier problem* in the literature [18]. Here, a set $\mathcal{P}$ of n clients (customer sites) and a set $\mathcal{Q}$ of m facilities (supplier sites) are given. The objective is to open a set $\mathcal{Q}_{opt} \subseteq \mathcal{Q}$ of k facilities such that the maximum distance of a client to its nearest facility from $\mathcal{Q}_{opt}$ is minimized. The k-supplier problem has numerous applications including facility location (e.g. placing k hospitals at some of the specified locations such that the maximum distance from any house to its nearest hospital is minimized), information retrieval, data mining etc.

The k-supplier problem is very relevant to sensors networks as it reflects the coverage issues in wireless sensor networks. In sensor networks, certain coverage

---

Wireless & Geometry.

M. Chrobak et al. (Eds.): ALGOSENSORS 2016, LNCS 10050, pp. 129–140, 2017.
DOI: 10.1007/978-3-319-53058-1_9

problems can be formulated as the k-supplier problem. Consider the placement of base stations or sensors to cover all the target points. We need to position exactly k base stations or sensors from a set of candidate locations such that all the target points are as close as possible to at least one of the base stations or sensors. In this scenario, optimal locations for the base stations or sensors would reduce the cost since the smaller range implies less power consumption of the batteries used in the sensors or base stations. Thus, a better algorithm for the k-supplier problem provides a better solution for the sensor deployment problem. The k-supplier problem has applications in several other facility location problems.

In Subsect. 1.1, we discuss the related works. In Sect. 2, we propose a parameterized 2-factor approximation algorithm for the k-supplier problem. In other words, our algorithm produces a subset $Q' \subseteq Q$ of size at most k such that the congruent disks of radius at most $2r_{opt}$ and centered at the points in Q' can cover all the points in $\mathcal{P}$, where r_{opt} is the radius of the disks in the optimum solution. Next, we present a heuristic algorithm based on Voronoi diagram, and justify its performance by an experimental study in Sect. 3. Finally, we conclude in Sect. 4.

1.1 Related Work

Hochbaum and Shmoys [12] and Gonzalez [10] provided 2-factor approximation algorithms in $O(n^2 \log n)$ and $O(nk)$ time respectively, for the k-center problem under general metrics. This is the best possible approximation bound as it is NP-hard to approximate beyond a factor of 2 for the k-center problem under general metrics [11]. However, Feder and Greene [9] gave a 2-factor approximation algorithm in $O(n \log k)$ time for the Euclidean k-center problem and showed that in the Euclidean metrics, this problem can not be approximated to within a factor of $\sqrt{3} \approx 1.73$ unless P = NP. Agarwal and Procopiuc [1] proposed an $(1 + \epsilon)$-factor approximation algorithm that runs in $O(n \log k + (\frac{k}{\epsilon})^{\sqrt{k}})$ time.

A generalization of the k-center problem, called the k-supplier problem, is available in the literature, where the set of given points is partitioned into two subsets Q and P, called facilities and clients respectively, and the objective is to choose k facilities such that the maximum distance of any client to its nearest chosen facility is minimum. In general metric, Hochbaum and Shmoys [13] gave 3-factor approximation algorithm in $O((n^2 + mn) \log(mn))$ time and proved that $(3 - \epsilon)$-factor approximation algorithm in polynomial time is not possible unless $P = NP$. However, in the Euclidean metric, Feder and Greene [9] gave 3-factor approximation algorithm for the Euclidean k-supplier problem with running time $O((n + m) \log k)$. They also showed that it is NP-hard to approximate the Euclidean k-supplier problem less than a factor of $\sqrt{7} \approx 2.64$. Furthermore, for fixed k, Hwang et al. [15] presented a $m^{O(\sqrt{k})}$ time algorithm for the Euclidean k-supplier problem in $\mathbb{R}^2$. Later, Agarwal and Procopiuc [1] gave $m^{O(k^{1-1/d})}$-time algorithm for the points in $\mathbb{R}^d$. Recently, Nagarajan et al. [18] gave a 2.74-factor approximation algorithm for the Euclidean k-supplier problem in any constant dimension with running time $O(mn \log(mn))$.

Various constrained versions of the k-center problem have been studied extensively in the literature. Hurtado et al. [14] considered Euclidean 1-center problem where the center is constrained to satisfy m linear constraints, and proposed an $O(n + m)$ time algorithm for it. Bose and Toussaint [5] provided an $O((n+m)\log(n+m))$ time algorithm for the 1-center problem where the disk is centered on the boundary of a convex polygon with m vertices, and the objective is to cover n demand points that may lie in inside or outside of the polygon. Brass et al. [4] studied a similar version of the k-center problem, where the points are in $I\!R^2$ and the centers are constrained to lie on a given straight line. It uses parametric search, and runs in $O(n \log^2 n)$ time. Karmakar et al. [16] proposed three algorithms for this problem with time complexities $O(nk \log n)$, $O(nk + k^2 \log^3 n)$ and $O(n \log n + k \log^4 n)$ respectively. Using parametric search, Kim and Shin [17] solved the 2-center problem for a given polygon with n vertices in $O(n \log^2 n)$ time, where the two centers are restricted to be at some vertices of the polygon, and the objective is to cover the entire polygon. Das et al. [6] provided $(1 + \epsilon)$-factor approximation algorithm for the k-center problem of a convex polygon with n vertices, where the centers are restricted to lie on a specified edge of the polygon. If the centers are restricted to be on the boundary of a convex polygon, Das et al. [6] presented an $O(n^2)$ time algorithm for the k-center problem, where $k = 1, 2$. In the same paper, they discussed a heuristic algorithm for the same problem, for $k \geq 3$. Later, Roy et al. [19] improved the time complexities of the same problem for $k = 1, 2$ to $O(n)$. Du and Xu [7] studied k-center problem for a convex polygon where the centers are restricted to lie anywhere on the boundary of the polygon and presented 1.8841-factor approximation algorithm, which runs in $O(nk)$ time, where n is the number of vertices of the polygon. Recently, Basappa et al. [3] gave $(1 + \frac{7}{k} + \frac{7\epsilon}{k} + \epsilon)$-factor approximation algorithm with running time $O(n(n + k)(|\log r_{opt}| + \log\lceil\frac{1}{\epsilon}\rceil))$ for this problem, where r_{opt} is the radius of disks in the optimal solution[1], $k \geq 7$, and $\epsilon > 0$ is any given constant.

Recently, Dumitrescu and Jiang [8] studied the following variation of the *constrained k-center* problem: given a set $\mathcal{P} = \{p_1, p_2, \ldots, p_n\}$ of n black points and a set $\mathcal{Q} = \{q_1, q_2, \ldots, q_k\}$ of k red points in $I\!R^2$, the goal is to find a set $\mathcal{D} = \{D_1, D_2, \ldots, D_k\}$ of k disks such that (i) the disk D_j must contain the red point $q_j \in \mathcal{Q}$ for $1 \leq j \leq k$, (ii) all points in $\mathcal{P}$ are covered by the union of the disks in $\mathcal{D}$ and (iii) the maximum radius of the disks in $\mathcal{D}$ is minimized. Dumitrescu and Jiang [8] showed that their *constrained k-center* problem is NP-hard and can not be approximated within 1.8279 factor. They proposed FPT-algorithms with 1.87, 1.71, and 1.61-factor approximation results in $O(3^k kn)$, $O(4^k kn)$, and $O(5^k kn)$ time respectively. Based on the generalization of the idea used for developing the above FPT-algorithms, they proposed $(1+\epsilon)$-factor approximation algorithm in $O(\epsilon^{-2k}n)$ time, where $\epsilon > 0$.

---

[1] We have taken $|\log r_{opt}|$ in the time complexity since r_{opt} may be less than 1.

2   FPT Algorithm for 2-Factor Approximation Result

2.1   Terminologies

Let $\mathcal{P} = \{p_1, p_2, \ldots, p_n\}$ and $\mathcal{Q} = \{q_1, q_2, \ldots, q_m\}$ denote a set of n clients and a set of m facilities respectively in $\mathbb{R}^2$. Throughout the paper we use $\delta(a, b)$ to denote the Euclidean distance between a pair of points $a, b \in \mathbb{R}^2$, and $\Delta(a, r)$ to denote the region covered by the disk of radius r centered at the point a. Let $D = \{\delta(p, q) \mid p \in \mathcal{P} \text{ and } q \in \mathcal{Q}\}$ be the set of distances between the points in $\mathcal{P}$ and the points in $\mathcal{Q}$. Let $r_1, r_2, \ldots, r_{mn}$ be the non-decreasing order of the members in D. Let $\mathcal{Q}_{opt}$ be an optimal solution of the k-supplier problem, and r_{opt} be the radius of the disks in $\mathcal{Q}_{opt}$.

Lemma 1. $r_{opt} \in \{r_1, r_2, \ldots, r_{mn}\}$.

Proof. Assume that $r_{opt} \notin \{r_1, r_2, \ldots, r_{mn}\}$. Then there must exist an i such that $r_i < r_{opt} < r_{i+1}$. Thus, no point in $\mathcal{P}$ lies on the boundary of any of the disks in the optimal solution centered at k points in $\mathcal{Q}$. This implies that we can reduce the radius of every disk and still cover $\mathcal{P}$. This contradicts the fact that r_{opt} is the minimum radius.   □

2.2   Approximation Algorithm

In this subsection we propose a parameterized 2-factor approximation algorithm for the k-supplier problem. The objective is to choose a subset $\hat{Q} \subseteq \mathcal{Q}$ of size at most k such that the union of k disks of radius $r \leq 2r_{opt}$ centered at the points in $\hat{Q}$ covers all the points in $\mathcal{P}$.

Let us first consider the following decision problem.

– *For a given radius r, does there exist a subset $Q' \subseteq \mathcal{Q}$ of size at most k such that the union of k disks of radius $2r$ centered at the points of Q' covers all the points in $\mathcal{P}$?*

We show that the above decision problem can be solved with time complexity $O(\alpha^k \text{polynomial}(m, n))$, where α is a predefined constant. For a given radius r, if the answer is positive, then it reports the chosen subset Q', where $|Q'| \leq k$. For a negative reply, an arbitrary subset of $k + 1$ points is reported.

We apply binary search in the set $D = \{r_1, r_2, \ldots, r_{mn}\}$ to find the minimum r for which the above decision problem returns a positive reply (a subset $Q' \subseteq \mathcal{Q}$, where $|Q'| \leq k$).

Let the point $p \in \mathcal{P}$ be covered by a disk of radius r centered at $q \in \mathcal{Q}$. Thus $q \in \Delta(p, r)$. Let us draw six radii in the circular region $\Delta(p, r)$ such that each pair of consecutive radii make an angle $\frac{\pi}{3}$ at the point p. These split the region $\Delta(p, r)$ into six equal sectors $\Delta^1, \Delta^2, \ldots, \Delta^6$ as shown in Fig. 1.

Lemma 2. *If $q \in \Delta^i, i \in \{1, 2, \ldots, 6\}$, then any disk of radius $2r$ centered at any point in the region Δ^i covers all the points of $\mathcal{P}$ that are covered by the disk of radius r centered at q.*

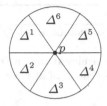

Fig. 1. Partitioning the disk $\Delta(p, r)$ into six equal sectors $\Delta^1, \Delta^2, \ldots, \Delta^6$

Proof. Follows from the triangle inequality and the facts that (i) $\delta(p', q) \leq r$ for any point $p' \in \mathcal{P}$ that is covered by the disk of radius r centered at q, and (ii) $\delta(q, q') \leq r$ for any point $q' \in \mathcal{Q}$ lying in Δ^i.   □

Algorithm 1. k-supplier($\mathcal{P}, \mathcal{Q}, k$)

1: **Input:** Two sets of points $\mathcal{P} = \{p_1, p_2, \ldots, p_n\}$ and $\mathcal{Q} = \{q_1, q_2, \ldots, q_m\}$, and a positive integer k.

2: **Output:** a set of points $\hat{Q} \subseteq \mathcal{Q}$ of size at most k and a radius $\hat{r}$ such that the union of the disks with radius $2\hat{r}$ centered at the points in $\hat{Q}$ covers all points in $\mathcal{P}$.

3: Let $\{r_1, r_2, \ldots, r_{mn}\}$ be the elements of $\{\delta(p, q) \mid p \in \mathcal{P} \ \& \ q \in \mathcal{Q}\}$ in non-decreasing order.

4: Set $\ell \leftarrow 1; \ h \leftarrow mn$;

5: **while** ($\ell < h$) **do**

6:     $mid = \lceil \frac{\ell+h}{2} \rceil$; $P' \leftarrow \mathcal{P}$; $Q' \leftarrow \emptyset$;/* Q' will hold at most k centers */

7:     $(flag, Q') = \mathbf{Cover}(P', \mathcal{Q}, k, r_{mid})$

8:     **if** ($flag$) **then**

9:        $\hat{Q} = Q'$; $\hat{r} = r_{mid}$; $h = mid - 1$;

10:    **else**

11:       $\ell = mid + 1$

12:    **end if**

13: **end while**

14: **return** $(\hat{Q}, \hat{r})$

We solve the decision problem for a given radius r as follows. We initialize $P' = \mathcal{P}$ and $Q' = \emptyset$. Choose a point $p \in P'$, and arbitrarily partition the disk $\Delta_1 = \Delta(p, r)$ into six equal sectors $\Delta_1^1, \Delta_1^2, \ldots, \Delta_1^6$ as shown in Fig. 1.

We consider each sector $\Delta_1^i, i = 1, 2, \ldots, 6$ separately. For each sector Δ_1^i, if $\mathcal{Q} \cap \Delta_1^i \neq \emptyset$, then we can choose any point $q \in \mathcal{Q} \cap \Delta_1^i$ (by Lemma 2). Update Q' by $Q' \cup \{q\}$. Let $R \subseteq \mathcal{P}$ be the set of points lying in $\Delta(q, 2r)$. We update $P' = P' \setminus R$. If the updated $P' \neq \emptyset$, we repeat the same process recursively to find $q' \in \mathcal{Q}$ by arbitrarily choosing a point $p' \in P'$ (updated), drawing $\Delta_2^i, i = 1, 2, \ldots, 6$, and then processing each Δ_2^i to update $Q' = Q' \cup \{q'\}$. The process along a path of the recursion stops if either the updated P' up to that level of recursion is empty, or the level of recursion is k.

Algorithm 2. Cover(P, Q, k, r)

1: **Input:** The set P of uncovered points, a set Q of m facility points, a positive integer k, and a radius r.

2: **Output:** *true* if cover of P is achieved with at most k disks of radius $2r$; *false* otherwise and a set Q' to hold centers of at most k disks.

3: **if** ($P = \emptyset$) **then**

4:    **return** (*true*, $\emptyset$)

5: **else if** ($k = 0$) **then**

6:    **return** (*false*, $\emptyset$)

7: **else**

8:    Consider the disk $\Delta(p, r)$ centered at an arbitrary point $p \in P$, and

9:    partition $\Delta(p, r)$ into six equal sectors $\Delta^1, \Delta^2, \dots, \Delta^6$ as in Fig. 1.

10:    Set $i \leftarrow 1$; $flag \leftarrow false$; $Q' \leftarrow \emptyset$

11:    **while** ($i \leq 6$ and *flag=false*) **do**

12:       **if** ($Q \cap \Delta^i \neq \emptyset$) **then**

13:          choose a point $q \in Q \cap \Delta^i$

14:          $P' = P \backslash (P \cap \Delta(q, 2r))$;

15:          ($flag, Q'$) = **cover**($P', Q, k - 1, r$)

16:          **if** (*flag=true*) **then**

17:             $Q' = Q' \cup \{q\}$

18:          **end if**

19:       **end if**

20:       Set $i = i + 1$;

21:    **end while**

22: **end if**

23: **return** (*flag*, Q')

Thus, our search process progresses in a tree like fashion, where the degree of each node in this search tree is at most six, and the maximum length from the root up to a leaf in any search path is at most k. At the end of this process, if the resulting set $P' = \emptyset$ along any one of the search paths explored, then we return the corresponding Q', otherwise we return Q' with an arbitrary subset of $k + 1$ points. Thus after executing this decision algorithm, it indicates a positive answer if $|Q'| \leq k$, and a negative answer if $|Q'| = k + 1$.

The pseudocode of the procedure for computing the minimum r such that at most k disks of radius $2r$ centered at the points in Q covers all the points in P is given in Algorithm 1.

Lemma 3. *If Algorithm 2 is invoked with $r = r_{opt}$, then it produces a positive reply. In other words, it produces a subset $Q' \subseteq Q$ of size at most k such that union of the disks with radius $2r$ centered at the points in Q' covers all the points in $\mathcal{P}$.*

Proof. Note that, in the optimum solution, each member of $\mathcal{P}$ is covered by some element of Q_{opt}. In Algorithm 2, suppose we have chosen some element $p_1 \in \mathcal{P}$, and its corresponding covering element $q_1 \in Q_{opt}$ lies in Δ_1^i of $\Delta(p_1, r_{opt})$. Let $P_1 = \mathcal{P} \backslash (\mathcal{P} \cap \Delta(q_1, r_{opt}))$. Now, in the 6-way search tree of Algorithm 2, if we proceed towards a point q_1' in the sector of $\Delta(p_1, r_{opt})$ containing q_1, then the points that remain uncovered are $P_1' = \mathcal{P}\backslash(\mathcal{P}\cap\Delta(q_1', 2r_{opt})) \subseteq P_1$ (see Lemma 2). Again, for any element $p_2 \in P_1'$, if its covering element is $q_2 \in \mathcal{Q}_{opt}$, and in the 6-way search tree if we proceed with a point q_2' in the sector of $\Delta(p_2, r_{opt})$ containing q_2, the remaining uncovered points $P_2' = \mathcal{P}\backslash(\mathcal{P} \cap (\Delta(q_1', 2r_{opt}) \cup \Delta(q_2', 2r_{opt})))$ will be a subset of $P_2 = \mathcal{P}\backslash(\mathcal{P} \cap (\Delta(q_1, r_{opt}) \cup \Delta(q_2, r_{opt})))$ (see Lemma 2). The maximum depth along such a search path will be less than or equal to k since $|\mathcal{Q}_{opt}| \leq k$. As we are exploring all possible search paths, the result follows. □

Lemma 4. *If $\hat{Q}$ and r are the output of Algorithm 1, then the union of the disks with radius $2r$ centered at the points of $\hat{Q}$ covers all the points in $\mathcal{P}$ and $r \leq r_{opt}$.*

Proof. Let $\mathcal{P}, \mathcal{Q}, k, r$ be the input of Algorithm 2. Lemma 3 says that if $r = r_{opt}$ then Algorithm 2 returns a subset $Q' \subseteq \mathcal{Q}$ such that the union of the disks with radius $2r$ centered at the points in Q' covers all the points in $\mathcal{P}$. Lemma 1 says that $r_{opt} \in \{r_1, r_2, \ldots, r_{mn}\}$. Algorithm 1 finds a minimum value $r \in \{r_1, r_2, \ldots, r_{mn}\}$ such that with the input $\mathcal{P}, \mathcal{Q}, k, r$, Algorithm 2 returns a set $Q'(=\hat{Q}) \subseteq \mathcal{Q}$ such that the union of the disks with radius $2r$ centered at the points in Q' covers all the points in $\mathcal{P}$. Therefore $r \leq r_{opt}$. □

Theorem 1. *Algorithm 1 for the k-supplier problem produces a 2-factor approximation result, and it runs in $O(6^k(n + m)\log(nm))$ time.*

Proof. Approximation factor follows from Lemma 4.

Let $T(n, m, k)$ be the running time of Algorithm 2, where $n = |\mathcal{P}|$, $m = |\mathcal{Q}|$. Since each node of the recursion tree has degree at most 6, and execution at that node takes $O(n+m)$ time, we have $T(n, m, k) = 6((m+n)+T(n, m, k-1))$. Again, since the depth of the recursion tree is at most k, we have the running time of the procedure **cover** (Algorithm 2) is $O(6^k(n+m))$. Algorithm 1 invokes Algorithm 2 at most $\log(mn)$ times. Thus the total running time is $O(6^k(n + m)\log(mn))$. □

Corollary 1. *For $k = O(\log n)$, the k-supplier problem in $I\!\!R^2$ has a 2-factor approximation algorithm that runs in polynomial time.*

We can extend the idea of solving the k-supplier problem in $I\!\!R^2$ to solve the k-supplier problem in $I\!\!R^3$. Here, we consider a ball of radius r in $I\!\!R^3$ instead of a disk of radius r in $I\!\!R^2$. We partition the ball of radius r into 12 equal sectors such that the distance between any two points in a sector is at most $\sqrt{2}r$. Remaining part of the algorithm for $I\!\!R^3$ is exactly the same as in the case of $I\!\!R^2$. Thus we have the following theorem.

Theorem 2. *For the k-supplier problem in $I\!\!R^3$, we can get a $(1 + \sqrt{2})$-factor approximation result in $O(12^k(n + m)\log(mn))$ time.*

Note: It needs to be mentioned that in Algorithm 2, instead of splitting the disk $\Delta(p, r)$ into 6 parts, and choosing points q in those parts, if we choose the point p itself and reduce the set of uncovered points by $\Delta(p, 2r)$, then we can generate a 2-factor approximation result for the unconstrained (general) version of the k-center problem in time $O(kn + n^2 \log n)$ time.

3   Heuristic Algorithm for the k-Supplier Problem in $I\!\!R^2$

In this section, we present a heuristic algorithm for the k-supplier problem based on Voronoi diagram [2]. Initially, we pick a set of k arbitrary points $Q' = \{q'_1, q'_2, \ldots, q'_k\} \subseteq Q$. We compute Voronoi diagram $VOR(Q')$ of the points in Q'. This forms the clusters of the points in $\mathcal{P}$, namely $\mathcal{P}_i = \{p \in \mathcal{P} \mid p \in vor(q'_i)\}$, where $vor(q'_i)$ is the Voronoi cell of the point $q'_i \in Q'$, $i = 1, 2, \ldots, k$. For each cluster $\mathcal{P}_i$, let d_i be the smallest disk centered at a point in Q such that $\mathcal{P}_i \subset d_i$. Let $r = \max\{r_1, r_2, \ldots, r_k\}$, where r_i is the radius of the disk d_i, and $Q'' = \{q''_i \mid q''_i$ is the center of the disk d_i, $i = 1, 2, \ldots, k\}$. We repeat this process by setting $Q' = Q''$. The process continues as long as the radius r decreases. The detailed pseudocode of this procedure is given in Algorithm 3.

Lemma 5. *At each iteration of Algorithm 3 the value of r never increases.*

Proof. Let $q_1, q_2, \ldots, q_k$ be the Voronoi sites (cluster centers) at the beginning of an iteration. The minimum enclosing disks of $\mathcal{P} \cap vor(q_1), \mathcal{P} \cap vor(q_2), \ldots, \mathcal{P} \cap vor(q_k)$ are $d'_1, d'_2, \ldots, d'_k$. Let the center of the disk d'_i be q'_i and the radius be r'_i, $i = 1, 2, \ldots, k$.

Now consider an arbitrary Voronoi cell $vor(q_i)$ $(1 \leq i \leq k)$. Without loss of generality assume that $vor(q_1), vor(q_2), \ldots, vor(q_t)$ are the neighbouring Voronoi cells of $vor(q_i)$. We consider the following cases: (a) $r'_i \geq \max\{r'_1, r'_2, \ldots, r'_t\}$, and (b) $r'_i < \max\{r'_1, r'_2, \ldots, r'_t\}$.   □

In this iteration, let d''_i denote the minimum enclosing disk of the points in $\mathcal{P} \cap vor(q'_i)$, whose center is at a point $q''_i \in Q$, and the radius is r''_i, $i = 1, 2, \ldots, k$. Here, $d''_i \neq d'_i$ only if $\mathcal{P} \cap vor(q'_i) \neq \mathcal{P} \cap vor(q_i)$.

Case (a): It is sufficient to show that $r''_i \leq r'_i$. On contrary assume that $r''_i > r'_i$. Therefore there exists at least one point $p \in \mathcal{P} \cap vor(q_s)$ (for some $s \in \{1, 2, \ldots, t\}$) in the previous iteration, but $p \in \mathcal{P} \cap vor(q'_i)$ in this iteration. We choose p to be the one which is farthest from q'_i among the points which entered from some other Voronoi cell to $vor(q'_i)$ in this iteration. The Voronoi partitioning suggests that $\delta(q'_i, p) \leq \delta(q'_s, p)$.

Again, observe that $r''_i \leq \delta(q'_i, p)$ since r''_i is the radius of the minimum enclosing disk d''_i containing all the points $\mathcal{P} \cap vor(q'_i)$ and its center q''_i may be different from q'_i. Since q'_s is the center of the minimum enclosing disk (centered at a point in Q) for a point set containing p, we have $\delta(q'_s, p) \leq r'_s$. Again, $r'_s \leq r'_i$ by our assumption. Thus we have, $r''_i \leq r'_i$, which leads to a contradiction.

Case (b): The lemma is trivial if $r''_i \leq r'_i$. Therefore we assume that $r''_i > r'_i$. Thus, we have at least one point $p \in \mathcal{P} \cap vor(q_s)$ (for some $s \in \{1, 2, \ldots, t\}$) such that p moves to the cell $vor(q'_i)$ in this iteration. From the properties of Voronoi partition $\delta(q'_i, p) \leq \delta(q'_s, p)$. As in the former case, we have $r''_i \leq \delta(q'_i, p)$ and $r'_s \geq \delta(q'_s, p)$. Therefore, $r''_i \leq \delta(q'_i, p) \leq \delta(q'_s, p) \leq r'_s$ i.e., $r''_i \leq r'_s$. Thus, the lemma follows in this case also.   □

Algorithm 3. k-Supplier-Heuristic$(\mathcal{P}, \mathcal{Q}, k)$

1: **Input:** A set $\mathcal{P}$ of n points, a set $\mathcal{Q}$ of m points, and a positive integer k.
2: **Output:** A set $\mathcal{D}$ of k disks of radius r centered at k points of $\mathcal{Q}$ such that
   $\mathcal{P} \subseteq \cup_{d \in \mathcal{D}} d$.
3: $r_{old} \leftarrow \infty$; $r_{new} = \max\{\delta(p,q) | p \in \mathcal{P} \& q \in \mathcal{Q}\}$
4: Let $Q' = Q'' = \{q_1, q_2, \ldots, q_k\} \subseteq \mathcal{Q}$ be an arbitrary subset of k points, called
   cluster centers.
5: **while** ($r_{new} < r_{old}$) **do**
6:    Set $r_{old} = r_{new}$, $Q' = Q''$
7:    Compute $VOR(Q')$.
8:    (/* Compute the cluster $\mathcal{P}_i = \mathcal{P} \cap vor(q_i)$ */)
9:    **for** $i = 1, 2, \ldots, n$ **do**
10:     Apply point location with the point p_i in $VOR(Q')$ to assign it in appropriate
        cluster.
11:   **end for**
12:   **for** $i = 1, 2, \ldots, k$ **do**
13:     Compute the convex hull $CH(\mathcal{P}_i)$ of the points in $vor(q_i)$, and
        the furthest point Voronoi diagram $FVD(\mathcal{P}_i)$ of the vertices of $CH(\mathcal{P}_i)$.
14:   **end for**
15:   Create two arrays, namely $r[1, 2 \ldots, k]$ initialized with $[\infty, \infty, \ldots, \infty]$ and
        $Q''[1, 2, \ldots, k]$ to store the new *cluster centers.*
16:   **for** $j = 1, 2, \ldots, m$ **do**
17:     Find i such that $q_j \in vor(q_i)$
18:     Consult $FVD(\mathcal{P}_i)$ to find a vertex p of $CH(\mathcal{P}_i)$ which is furthest from q_j
        among the other vertices of $CH(\mathcal{P}_i)$.
19:     Compute $\delta(p, q_j)$;
20:     **if** $\delta(p, q_j) < r(\mathcal{P}_i)$ **then**
21:       Assign $r[i] = \delta(p, q_j)$; $Q''[i] = q_j$
22:     **end if**
23:   **end for**
24:   (/* For each i, the minimum enclosing disk d_i for the cluster $\mathcal{P}_i$ is centered at
        $Q''[i]$ and has radius $r[i]$. */)
25:   Compute $r_{new} = \max\{r[1], r[2], \ldots, r[k]\}$
26: **end while**
27: Set $r = r_{old}$, $\mathcal{D} = \emptyset$
28: **for** $(i = 1, 2, \ldots, k)$ **do**
29:   Let d_i be the disk of radius r centered at $q_i \in Q'$
30:   $\mathcal{D} = \mathcal{D} \cup \{d_i\}$
31: **end for**
32: Return $(r, \mathcal{D})$

Lemma 6. *The worst case time complexity of every iteration in Algorithm 3 is* $O((m+n)\log(nk))$.

Proof. In the **while** loop, computing Voronoi diagram (line 7) for a set of k points takes $O(k \log k)$ time. All the clusters $\mathcal{P}_i$ $(i = 1, 2, \ldots, k)$ can be computed in $O(n \log k)$ time using planar point location in $VOR(Q')$. In order to compute the

Table 1. Radii of disks centered by our proposed heuristic algorithm and the algorithm by Nagarajan et al.'s [18], and the execution times of these two algorithms.

n	m	k	r_{vor}	r_{nag}	t_{vor}	t_{nag}
100	50	20	385.1651	767.8174	0.0204	0.0052
200	100	50	331.8947	474.1622	0.0536	0.0148
500	400	50	271.5503	354.4516	0.4824	0.0767
500	400	100	187.3902	330.0519	0.9357	0.0883
500	400	200	165.4990	275.2376	1.8368	0.0871
500	400	300	157.4919	352.6021	1.8220	0.0758
800	400	100	220.0354	348.7830	1.7462	0.1114
800	400	200	162.2155	299.2550	2.5599	0.1308
800	400	300	188.4023	431.0273	2.5415	0.1003
800	600	100	226.4105	315.7258	1.9460	0.1540
800	600	200	183.6834	347.9941	3.8523	0.1372
800	600	300	153.9939	287.3068	5.7032	0.1638
800	600	400	162.6617	417.0634	5.0574	0.1249
800	600	500	129.5877	290.8249	6.2892	0.1639
800	700	100	279.8605	313.5049	1.5080	0.1685
800	700	200	168.2815	248.6119	4.4462	0.2011
800	700	300	145.6643	279.1926	6.6276	0.1942
800	700	400	134.3790	337.201	5.9084	0.1706
800	700	500	139.7529	224.3580	7.5274	0.2149
800	700	600	156.6618	239.7690	8.7611	0.2089
1000	800	100	190.4600	243.5794	4.1017	0.2829
1000	800	200	158.9247	256.8944	6.0269	0.2598
1000	800	300	186.3586	253.9525	9.1140	0.2611
1000	800	400	120.5850	246.3848	11.9620	0.2714
1000	800	500	113.2485	228.0754	10.0219	0.2947
1000	800	600	134.3641	266.7492	11.9123	0.2483
1000	800	700	100.2370	248.2916	14.0810	0.2705
1000	900	100	226.1000	300.7388	4.6818	0.2506
1000	900	200	143.3740	252.6710	9.0590	0.2932
1000	900	300	133.1333	229.2618	10.1513	0.3144
1000	900	400	122.3652	293.7928	13.5053	0.2568
1000	900	500	113.3674	238.9505	16.8278	0.3009
1000	900	600	108.2268	261.4600	18.1648	0.2833
1000	900	700	101.3077	217.5035	20.4861	0.3194
1000	900	800	100.5504	230.9309	22.9282	0.3095

new cluster centers, for each point $q \in Q$ first identify in which Voronoi cell it falls (the **for** loop in line 16), and then locate its furthest neighbor (**for** loop in line 18) among the vertices of the convex hull of $\mathcal{P}_i$. This needs $O(m(\log k + \log n_i))$ time, where $n_i = |CH(\mathcal{P}_i)|$. The computation of $CH(\mathcal{P}_i)$ and $FVD(\mathcal{P}_i)$ for all $i = 1, 2, \ldots, k$ need $O(n \log n)$ time. Thus, the overall time complexity for a single iteration is $O((m + n) \log(nk))$. □

3.1 Experimental Results

We implemented our Voronoi diagram based heuristic algorithm, and the best known algorithm available in the literature for k-supplier problem in $I\!R^2$ [18]. We have used Matlab on a machine equipped with hardware configuration of Intel® Pentium(R) CPU G870 @ 3.10 GHz × 2, 1.8 GiB RAM and running 64-bit Ubuntu 12.03 to implement both the algorithms. We have chosen two sets of points within a square 20 times for different values of m, n and k. We executed the algorithms for each chosen instance and finally returned average values as output (see Table 1). Radii of disks for our Voronoi diagram based heuristic algorithm and for the algorithm in [18] are denoted by r_{vor} and r_{nag} respectively and the execution times (in second) of the algorithms are denoted as t_{vor} and t_{nag} respectively. The computed radii results indicate that our Voronoi diagram based heuristic algorithm produces much better result than the algorithm in [18]. However, the execution time of our algorithm is little more than that of [18].

4 Conclusion

In this paper we have proposed a fixed parameter tractable (FPT) algorithm for k-supplier problem in $I\!R^2$. Our proposed FPT algorithm produces 2-factor approximation result. The running time of the proposed FPT algorithm is $O(6^k(n + m) \log(nm))$, where k is the parameter. The proposed FPT algorithm can be extended to $I\!R^3$ easily, and we can get a $(1 + \sqrt{2})$-factor approximation result with running time $O(12^k(n + m) \log(nm))$. For k-supplier problem in $I\!R^2$, we have also proposed a heuristic algorithm based on Voronoi diagram. We did experimental study on our heuristic algorithm and also the best known algorithm available in the literature [18] for the k-supplier problem in $I\!R^2$. Experimental results indicate that our heuristic algorithm performs better than the algorithm available in [18] with very minor degradation in the running time.

References

1. Agarwal, P.K., Procopiuc, C.: Exact and approximation algorithms for clustering. Algorithmica **33**, 201–226 (2002)
2. de Berg, M., Cheong, O., Van Kreveld, M., Overmars, M.: Computational Geometry Algorithms and Applications. Springer, Heidelberg (2008)
3. Basappa, M., Jallu, R.K., Das, G.K.: Constrained k-center problem on a convex polygon. In: Gervasi, O., et al. (eds.) ICCSA 2015. LNCS, vol. 9156, pp. 209–222. Springer, Heidelberg (2015). doi:10.1007/978-3-319-21407-8_16

4. Brass, P., Knauer, C., Na, H.S., Shin, C.S.: Computing k-centers on a line. CoRR abs/0902.3282 (2009)
5. Bose, P., Toussaint, G.: Computing the constrained Euclidean, geodesic and link center of a simple polygon with applications. In: Proceedings of Pacific Graphics International, pp. 102–112 (1996)
6. Das, G.K., Roy, S., Das, S., Nandy, S.C.: Variations of base station placement problem on the boundary of a convex region. Int. J. Found. Comput. Sci. **19**(2), 405–427 (2008)
7. Du, H., Xu, Y.: An approximation algorithm for k-center problem on a convex polygon. J. Comb. Opt. **27**(3), 504–518 (2014)
8. Dumitrescu, A., Jiang, M.: Constrained k-center and movement to independence. Discrete Appl. Math. **159**(8), 859–865 (2011)
9. Feder, T., Greene, D.: Optimal algorithms for approximate clustering. In: Proceedings of the 20th ACM Symposium on Theory of Computing, pp. 434–444 (1988)
10. Gonzalez, T.F.: Clustering to minimize the maximum intercluster distance. Theor. Comput. Sci. **38**, 293–306 (1985)
11. Hochbaum, D.: Approximation Algorithms for NP-Hard Problems. PWS Publishing Company, Boston (1995)
12. Hochbaum, D.S., Shmoys, D.B.: A best possible heuristic for the k-center problem. Math. Oper. Res. **10**(2), 180–184 (1985)
13. Hochbaum, D.S., Shmoys, D.B.: A unified approach to approximation algorithms for bottleneck problems. J. ACM **33**(3), 533–550 (1986)
14. Hurtado, F., Sacriscan, V., Toussaint, G.: Facility location problems with constraints. Stud. Locat. Anal. **15**, 17–35 (2000)
15. Hwang, R., Lee, R., Chang, R.: The generalized searching over separators strategy to solve some NP-hard problems in subexponential time. Algorithmica **9**, 398–423 (1993)
16. Karmakar, A., Das, S., Nandy, S.C., Bhattacharya, B.K.: Some variations on constrained minimum enclosing circle problem. J. Comb. Opt. **25**(2), 176–190 (2013)
17. Kim, S.K., Shin, C.-S.: Efficient algorithms for two-center problems for a convex polygon. In: Du, D.-Z.-Z., Eades, P., Estivill-Castro, V., Lin, X., Sharma, A. (eds.) COCOON 2000. LNCS, vol. 1858, pp. 299–309. Springer, Heidelberg (2000). doi:10.1007/3-540-44968-X_30
18. Nagarajan, V., Schieber, B., Shachnai, H.: The Euclidean k-supplier problem. In: Goemans, M., Correa, J. (eds.) IPCO 2013. LNCS, vol. 7801, pp. 290–301. Springer, Heidelberg (2013). doi:10.1007/978-3-642-36694-9_25
19. Roy, S., Bardhan, D., Das, S.: Base station placement on boundary of a convex polygon. J. Parallel Distrib. Comput. **68**, 265–273 (2008)

Author Index

Printed in the United States
By Bookmasters